Seeing Renaissance Glass

This book is part of the Peter Lang Humanities list.
Every volume is peer reviewed and meets
the highest quality standards for content and production.

PETER LANG
New York • Bern • Berlin
Brussels • Vienna • Oxford • Warsaw

Sarah M. Dillon

Seeing Renaissance Glass

Art, Optics, and Glass of Early Modern Italy, 1250–1425

PETER LANG
New York • Bern • Berlin
Brussels • Vienna • Oxford • Warsaw

Library of Congress Cataloging-in-Publication Data

Names: Dillon, Sarah M., author.
Title: Seeing Renaissance glass: art, optics, and glass of
early modern Italy, 1250–1425 / Sarah M. Dillon.
Description: New York: Peter Lang, 2018.
Includes bibliographical references and index.
Identifiers: LCCN 2018027569 | ISBN 978-1-4331-4834-7 (hardback: alk. paper)
ISBN 978-1-4331-4835-4 (ebook pdf) | ISBN 978-1-4331-4836-1 (epub)
ISBN 978-1-4331-4837-8 (mobi)
Subjects: LCSH: Optics and art—Italy—History—To 1500. | Glass.
Glass art—History. | Italy—Civilization—476–1268.
Italy—Civilization—1268–1559.
Classification: LCC N72.O68 D55 2018 | DDC 748.0945—dc23
LC record available at https://lccn.loc.gov/2018027569
DOI 10.3726/b14280

Bibliographic information published by **Die Deutsche Nationalbibliothek**.
Die Deutsche Nationalbibliothek lists this publication in the "Deutsche
Nationalbibliografie"; detailed bibliographic data are available
on the Internet at http://dnb.d-nb.de/.

Cover image: *Illusionistic Niche with Liturgical Vessels*, ca. 1330, by Taddeo Gaddi.
Baroncelli Chapel, Santa Croce, Florence.
Source: Raffaello Bencini/Alinari Archives, Florence.

29 Broadway, 18th floor, New York, NY 10006
www.peterlang.com

For Henry, may you never tire of learning new things
by seeing things through the eyes of others.

Contents

Illustrations

Chapter 1

Chapter 2

Chapter 3

Chapter 4

Chapter 5

Chapter 6

Acknowledgments

I can still remember my first encounter with the book *Visuality Before and Beyond the Renaissance* edited by Robert S. Nelson. Since that day many years ago, the captivating—but impossibly futile—mission of seeing through the eyes of another person living long ago or far away has had an unrelenting hold on both my scholarship and pedagogy and, subsequently, it has informed this book in ways that go beyond the traditional footnote. It also inspired this project in another sense; because none of the essays therein specifically treated Italian art of the fourteenth century my questions and curiosity about this period increased. Studying at Syracuse University's Villa Rossa in Florence answered some of these questions because I was able to live with the art of the Renaissance in a sense, walking the streets that Dante and Giotto once had. But this experience raised other, more informed questions. I pursued these questions while completing my doctorate at The Graduate Center and on research trips to Italy. As this story suggests, this book would not have been possible without my friends and professors from Syracuse and The Graduate Center who helped shape my questions along the way, or otherwise supported my pursuit of them some, including Jim Saslow, Jenn Ball, Cynthia Hahn, Elinor Richter, Barbara Lane, Rachel Kousser, Gary Radke, Jonathan Nelson, Barbara Deimling, Rab Hatfield, Jillian Domenici, Emily Schiavone, Kate Mendillo, Jeremy Glatstein, Whitney Thompson, Andrea

Ortuno, Anne Vaugniaux, Patricia Rocco, Trinity Martinez, and Ellen Hurst. I also need to thank my colleagues at Kingsborough Community College for their continued support; the PSC-CUNY Research Award Program of the Research Foundation of CUNY; and the museums, individuals, and institutions that have dedicated their photographs of artworks to the public domain or otherwise made them accessible and shareable. Lastly it gives me great joy to thank my family, both immediate and extended, for shaping my ideas, outlooks, and successes. Without all of you, I would not be me. And most importantly to my husband, there are no words to express how much you inspire me and how grateful I am for your support. You make my scholarship stronger and my days brighter; you're the cream in my coffee.

1

Introduction to Seeing Renaissance Glass

Art, Optics, and Glass of Early Modern Italy, 1250–1425

This is an account of how the seemingly mundane medium of glass laid the foundation for many of the key developments associated with Italian Renaissance art and culture. It may seem counterintuitive, at first, to contend that a lowly, man-made material like glass could have contributed to an era that witnessed the rise of the notion of artist as divine genius. But as the case studies found throughout this book make clear, the medium of glass had certain attributes and associations that made it unique among all other available materials. A survey of the chapter titles reveals the first of these attributes: glass could assume many different visual forms, from stained glass to gilded glass to transparent glass. Each of these iterations had its own striking visual qualities as well as a host of important symbolic interpretations. Another attribute that contributed to glass's unique position in the early modern period is the process by which it was made. In an almost alchemical transformation, base ingredients such as plant ash and sand were combined to make a product that, in its final form, resembled precious crystal or gemstones. Third, and perhaps most importantly for this study, is the fact that early modern religious, scientific, and artistic circles were all engaging with glass simultaneously—either physical pieces of glass or the symbolism associated with glass objects—contributing to a "network of glass," that is, a web of interrelated activities engaged in some way with glass or glass-related ideas.

This network of glass will be explored in more detail throughout the following chapters but a brief description is useful here to illustrate the approaches and methodologies central to the book. By the end of the thirteenth century, glass was used to make everything from utilitarian drinking vessels to optical lenses to holy reliquaries. Natural scientists performed experiments with mirrors and lenses while church fathers contemplated the spiritual symbolism of glass and glass objects. At the same time, glass was being fashioned into products available to the layperson. Glass windows, mirrors, and—perhaps most revolutionary of all—eyeglasses, which were invented in Pisa around 1280, were all changing the optical experience of the average person on a very practical level.

Amid this complex interaction of glass's technological, scientific, and religious associations, artists increasingly incorporated glass into their commissions in particularly innovative ways, either as actual panels of glass or as depictions of glass objects. Not only did glass feature prominently in some of the period's most important works but analyses of these artworks suggest that the artists may have been consciously referencing the relationship between glass and the optical sense. Thus, while the connection between optical technology and the visual arts is commonly associated with seventeenth-century artists such as Vermeer, an examination of thirteenth- and fourteenth-century Italian artists reveals related phenomena occurring about four hundred years earlier. Therefore, as the following chapters demonstrate, artworks featuring glass can serve as important points of intersection among the period's scientific theories, religious beliefs, and artistic expression.

In addition to the network of glass outlined here, there were more direct connections between the medium of glass and the sense of sight. Etymological evidence suggests a relationship between glass and optics as early as antiquity; the ancient Latin term for glass, *vitrum,* derived from the verb *videre*, meaning "to see."[1] This etymological link between glass and vision would have been strengthened in the twelfth century when an influential Arabic treatise on ophthalmology, which describes the eye as glass-like, became widely available in the Latin West. There were also a host of vitreous technologies that were directly related to the sense of sight: glass windows, glass mirrors, and eyeglasses, among others. With such considerations in mind, new significance is given to the use of glass or glass paste for the eyes of late medieval or early Renaissance sculptures—as in the case of Arnolfo di Cambio's Madonna originally intended for the façade of the cathedral of Florence (Figure 1.1)—which may have both reflected and reinforced the relationship between the vitreous medium and the eye. But what exactly does this reappraisal of such artistic practices indicate? What did Arnolfo intend to convey

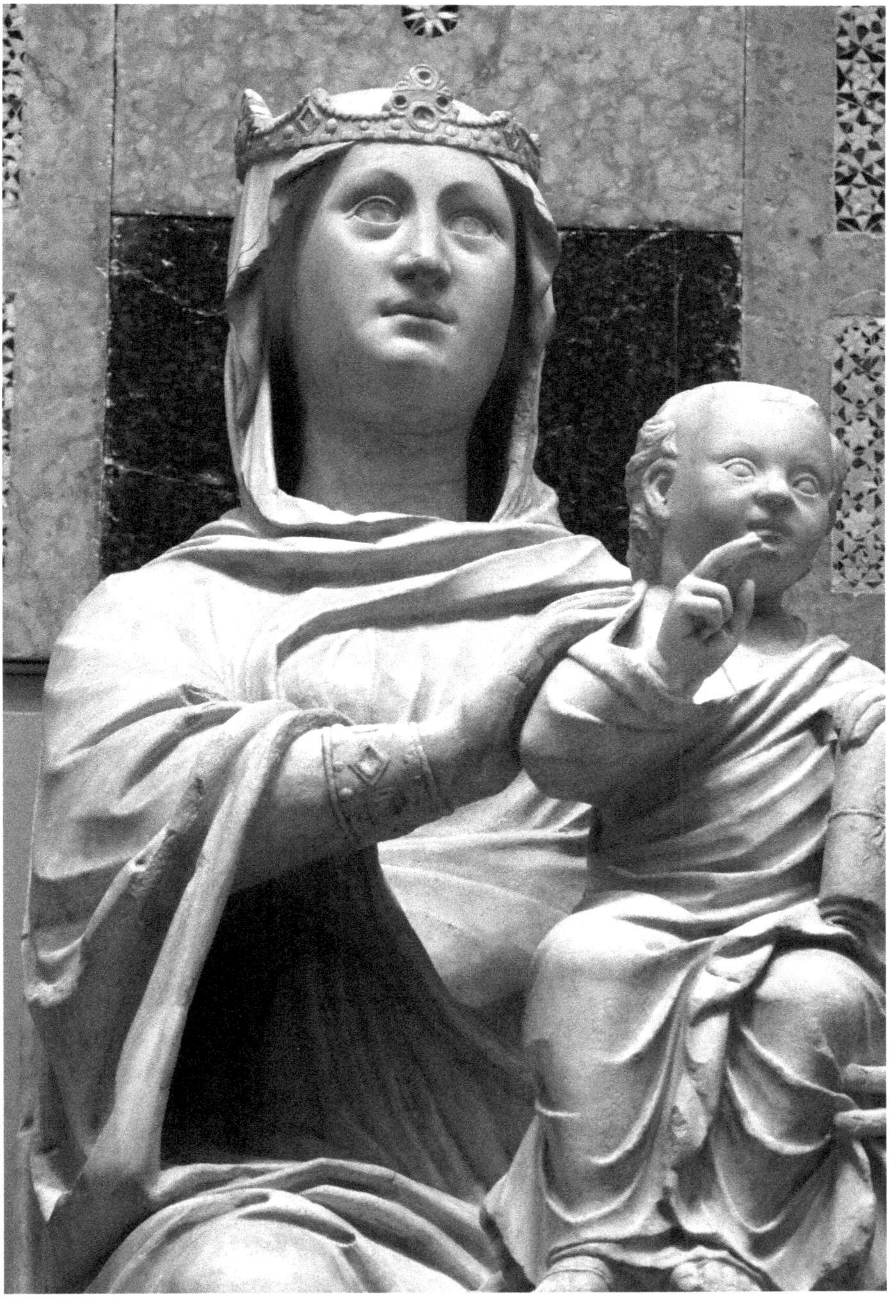

Figure 1.1: Arnolfo di Cambio, *Madonna and Child,* ca. 1296–1302, Museo dell'Opera del Duomo, Florence. Source: Francesco Bini via Wikimedia Commons (CC BY-SA).

when he used glass for the eyes of his Madonna? Was it simply a way to simulate the visual appearance of an actual human eye? Or was the artist trying to suggest a scientific understanding of the eye's anatomy as described in the Arabic treatise? To what degree was the artist reflecting optical theory, and to what degree was he perpetuating or disseminating it? These are the types of questions I ask about several different artists throughout the book.

Seeing Renaissance Glass is the first study to ask these types of questions about this particular period, namely the late-thirteenth through early-fifteenth centuries. There are several reasons why such an examination has not yet been conducted. As mentioned the seventeenth century is the era most often associated with the interplay of art and optics, and with good reason. Seventeenth-century artists in the Netherlands would have been aware of, and perhaps would even have had access to, optical instruments such as the camera obscura, microscope, and telescope. Another period typically associated with the intersection between art and science is the fifteenth century, specifically the mid to late part of the century, that is, the period after the development of linear perspective in 1425. Much has been written about the impact of linear perspective on both the artistic production of Renaissance art and, in turn, on wider cultural phenomena such as the period's visuality, technology, and scientific developments. Samuel Edgerton's work in particular has contributed greatly to an understanding of this dynamic interplay with his influential books *The Renaissance Rediscovery of Linear Perspective* and, more recently, *The Mirror, the Window, and the Telescope: How Renaissance Linear Perspective Changed Our Vision of the Universe*. For as he asserts, linear perspective "not only altered how we represent what we see but how we actually see a priori."[2] The present study applies similar lines of inquiry to the long fourteenth century, thereby expanding this debate to a new set of artists and artworks.

Another reason the duecento and trecento have been largely overlooked when it comes to such questions is because the period itself presents obstacles. Traditional art-historical narratives have tended to approach the fourteenth century in a bifurcated manner, dividing it at mid-century and treating the first half as pre-Black Death art and the second as post-Black Death art. There is no doubt that the 1348 bout of the bubonic plague had drastic effects on the lived experiences of the people in towns such as Siena, which lost an estimated 75 percent of its population. The impact on the artistic production and evolution of style, however, has been greatly debated ever since Millard Meiss originally published *Painting in Florence and Siena After the Black Death: The Arts, Religion and Society in the Mid-Fourteenth Century* in 1951. This ongoing dialog has shaped the historiography of trecento art to such a degree that it is difficult to find comprehensive studies of the period. Judith Steinhoff's work is an important exception. She finds continuous threads running throughout the entire century, allowing her to analyze the artistic evolution of these artists on their own terms, a process that yields fascinating insights into the period's appreciation of plurality and multiplicity. *Seeing Renaissance Glass* contributes to such discussions by adding new themes and case studies that unite pre- and post-plague Italian artistic production.

Another reason the scholarly literature on late medieval and early Renaissance art has not yet addressed the widespread use of glass by the artists of this time in a comprehensive manner is due to the artworks themselves and the ways they have been traditionally photographed. Ideally any study of these artworks involves careful, detailed, on-site observation, but in reality photographic reproductions of these artworks are oftentimes heavily utilized, and therefore it is important to consider that reproductions of the works might influence or shape our perception of them. When photographing artworks that incorporate glass panels, the reflections caused by glass tend to distract from or otherwise obscure stylistic details, which along with evidence of the hand of the artist, were the primary focus of traditional art-historical accounts of this period for most of its history. Thus, while many are aware of Simone Martini's *Maestà* in Siena for its combination of political and religious meaning, inventive composition, and portrayal of the Virgin as Heavenly Queen, a less studied aspect of the work is the fact that it has roughly twenty-five panels of gilded glass set into the wall within the depiction of the throne and halos of the Virgin and Child (Figure 1.2).[3] Yet as this book illustrates, the visual effects created by these small, perhaps unassuming panels, were prominent to contemporary viewers and may be central to fully understanding the artist's intentions.

Considering that many photographic reproductions of fourteenth-century art sought to essentially erase the presence of panels of glass, it should not be surprising that a comprehensive study of trecento artistic glass is still needed. However, this is not to say that all instances of glass in early Renaissance art have been ignored. Oftentimes conservation reports and exhibition catalogs record detailed information about the presence and condition of glass panels found in artworks, and in fact there have been several important studies of fourteenth-century objects made with glass or referencing glass. In terms of the study of mirrors, Herbert L. Kessler's 2011 piece in *Speculum* is most thorough and complete in both its consideration of the mirror's relationship to early modern visuality and scholarly bibliography.[4] However, no current studies simultaneously consider the many different facets of glass nor do they situate their case studies firmly within a network of glass, making *Seeing Renaissance Glass* unique in its cohesive study

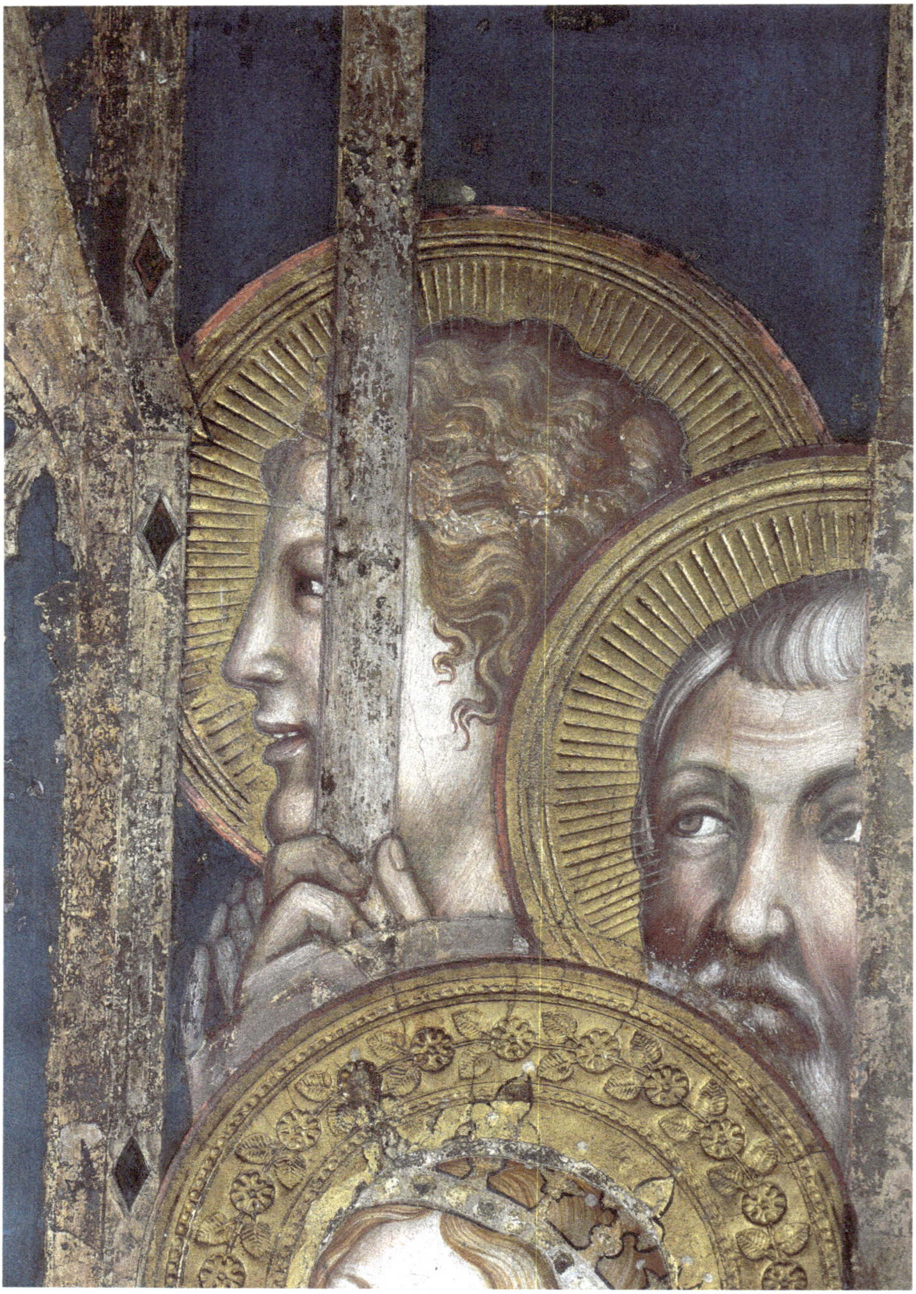

Figure 1.2: Simone Martini, Detail of *Maestà* showing diamond-shaped panels of glass, ca. 1315, Palazzo Pubblico, Siena. Source: Alinari Archives, Firenze.

of the various ways early modern Italian artists conceptualized, incorporated, and referenced the medium of glass in their work.

One reason it is important to investigate the relationship among early modern art, glass, and optical theory is because this period witnessed significant changes in all three fields. As will be discussed throughout this book, the art was generally moving towards a more naturalistic style, optical technologies made of glass were becoming more widely available, and there were new optical theories, such as those of Aristotle and Alhazen, to consider. The two dominant visual theories available during the late-medieval and early-Renaissance periods were extramission vision and intromission vision. In extramission vision, a theory promoted by Plato among others, the eye sent out fiery rays to the perceived object. On the other hand, intromission vision, the model mostly endorsed by Aristotle, described how the object emitted phantomlike reproductions of itself into the eye. Prior to the influx of original Islamic treatises and Arabic translations of ancient texts in the twelfth and thirteenth centuries, western Christian understanding of natural science in general, and optics more specifically, was derived from ancient Roman sources such as Seneca (ca. 4 BCE–65 CE), Pliny the Elder (ca. 23–79 CE), and Chalcidius (fl. early 4th c.), the last of whom transmitted the visual theories of Plato in his translation of the first part of *Timeaus.* These ideas were adapted to the Christian context through Augustine (354–430 CE), resulting in a modified version of Plato's original theories, but ones that had great influence on the medieval church from the fourth century through the end of the thirteenth century.[5] Aside from Chalcidius's version of Plato's *Timeaus*, however, the texts did not consist of complete optical studies. Rather, one needed to piece together various comments in order to compile information.

Departures from this mostly Platonic tradition were made by William of Conches (ca. 1080–1154), Adelard of Bath (fl. 1116–1142), and Robert Grosseteste. (ca. 1168–1253), despite the fact that these theorists considered themselves proponents of Plato.[6] William of Conches deviated from the extramission model when he described how the rays not only leave the eye but upon making contact with the object, they return to the beholder.[7] His focus on the return to the eye establishes a greater interest in the intromission position. Adelard, like William of Conches, also stressed the return of the visual information to the eye. Robert Grosseteste continued this trajectory. His theory is essentially Platonic but he, like Adelard, begins to reveal a debt to Arabic sources. Thus, with the influx and integration of Arabic theory, there was a growing interest in intromission and the incorporation of Arabic theories during the twelfth and thirteenth centuries.

The work of Albertus Magnus (ca. 1200–1280) and, to an even greater extent, Roger Bacon (ca. 1220–1292), mark a significant turning point in western optics. In his works *De anima* and *De sensu*, Albertus Magnus discussed and dismissed many of the previous theories of vision, including aspects of those put forth by Plato, Euclid, and Al-Kindi.[8] Instead he drew from Avicenna and Averroes to defend Aristotle's theory that vision occurs when the object alters the transparent medium between it and the eye and then the eye absorbs the altered medium.[9] It was Bacon, though, who pioneered the first Western visual theory to fully integrate and synthesize all previous optical knowledge, incorporating both ancient optical treatises, such as those by Aristotle, Euclid, and Ptolemy, as well as Arabic sources by Avicenna, Averroes, Al-Kindi, and Alhazen.[10]

It was specifically through studying Alhazen that Bacon formulated his mode of intromission, which was based on the fact that all objects issue species or rays in all directions in the form of a visual pyramid with its base at the perceived object and its apex at the observer's eye, with the stronger, perpendicular rays overpowering the weaker oblique ones to create a coherent view of the object.[11] In addition to the visual model of Alhazen, Bacon also adopted many of his requirements for the visual process to occur (i.e., light, size of the object, transparency of the intervening medium, the perceived object's density, the separation between the object and eye, and the time and health of the eye) as well as visual properties (i.e., light, shape, etc.).[12]

This theoretical background is important because it suggests the notion of vision was not static, and therefore any attempt to explore the relationship among art, optics, and glass must take this into account. In general there seems to have been a gradual shift from the extramission theory to intromission, but at least to some degree, there was continual debate. As late as the fifteenth century, Leonardo da Vinci questioned the role of visual rays, which are often associated with the extramission theory of vision.[13] Contextualizing the current study of art and glass within the context of optical theory is further important because, as previously mentioned, glass objects were used in optical experiments and therefore may have held certain scientific associations. For instance, it was through studying the effects visible in mirrors and lenses that early modern thinkers could directly observe the properties of light and sight.[14] In "Part Three of Perspective" of *Opus majus*, Bacon discussed vision in reflected and refracted lines and demonstrated his theories with mirrors.[15] Through observing the effects seen in mirrors, he was able to explain the various visual effects produced in different types of mirrors.[16]

Chapter Summaries

The chapters that follow each examine a different type of artistic glass—stained glass, gilded glass, and transparent glass—by analyzing several artworks that feature the type of glass in question and contextualizing these case studies within the period's optical technology, visual theory, and spiritual symbolism. Chapter 2 examines stained glass windows found in some of Italy's most famous chapels and cathedrals. Although stained glass is traditionally associated with the French Gothic period, Italian artists such as Duccio, Simone Martini, and Taddeo Gaddi designed stained glass windows and integrated them within larger artistic programs in ways that may be characteristic of a distinctly Italian approach. Furthermore, Italian artists who worked with stained glass oftentimes also made reference to other types of glass throughout their oeuvre. Simone Martini, to cite just one example, designed stained glass windows for his chapel of Saint Martin in the Lower Church at San Francesco in Assisi, depicted glass windows within his adjacent paintings, and he included gilded glass panels in his panel of Saint Louis of Toulouse and *Maestà* fresco (Figure 1.2).

Chapter 3 focuses on artists who incorporated pieces of gilded glass into their panel paintings, sculptures, and frescoes. Famous artworks such as Giotto's crucifix for Santa Maria Novella, Orcagna's tabernacle for Orsanmichele, and Simone Martini's *Maestà* for Siena's Palazzo Pubblico all employed actual panels of glass to enhance the visual appeal and religious significance of the works. This chapter also examines what appears to be a specific adaptation of gilded glass, *verre églomisé,* that is, the process of inscribing imagery into a gold-leaf-backed panel of glass. This chapter follows this medium's historical development and use in the late-thirteenth and fourteenth centuries by exploring the early modern development of this practice, beginning with its use by Nicola Pisano in the late-thirteenth century. It then situates the various uses of the medium within the context of artistic practices and visual theories and ultimately asserts that verre églomisé allowed the artist to combine naturalistic inscribed imagery with the powerful lighting effects of gilded glass in new ways. Or in other words, the medium of verre églomisé enabled artists to render images of holy figures composed almost entirely of symbolic golden light.

Both Chapters 2 and 3 find artists used glass to create lighting effects that resonated with contemporary religious beliefs and scientific theories about light and sight. Some of these ideas include Abbot Suger's concept of the *lux nova*, Bonaventure's theology of light, and Plato's extramission theory of vision. Through their analysis of a variety of case studies, these chapters reveal that Italian artists of the thirteenth and fourteenth centuries may have been incorporating glass in order to enhance the devotional efficacy of their artworks and were therefore engaged with theoretical principles and theological tenets in more complex ways than has previously been thought. Or, in other words, these artists were using glass in various ways to harness the power of one's earthly sight in service of divine insight.

Chapters 4 and 5 explore the development, dissemination, and uses of transparent or virtually colorless glass. Along with this revolutionary development in glassmaking and the new types of products it enabled came new ways of thinking about glass and its symbolic potential. Chapter 4 explains the importance of the medium's eastern origins and follows objects made from colorless glass as they made their way into the collections and the consciousness of the Latin West. After the 1204 Sack of Constantinople, precious Byzantine and Islamic glass and crystal objects infiltrated Europe, oftentimes as reliquaries either imported through trade or carried home by devout pilgrims as souvenirs from their trips to the Holy Land. Along with these glass and crystal objects, the Latin West also imported methods of glass production and optical treatises. This chapter considers the interrelated nature of these phenomena, and through exploration of the relationship between optical science, the growing interest in visible relics, and glass, it suggests that the medieval West was significantly shaped by the artistic and intellectual climate of the Islamic East.

A clearer understanding of the relationship among art, science, and religion in early modern Italy is revealed by juxtaposing early modern artworks featuring transparent glass panels or painted depictions of glass objects alongside contemporaneous scientific uses of glass and theories about glass. New types of glass coincided with new ideas about sight. For instance, as the use of transparent glass grew, there was a shift in intellectual circles about the nature of the visual process from extramission, which was associated with Plato, to intromission, which was more closely aligned with the ideas of Aristotle.

Chapter 5 outlines the ways in which the emerging visual theory correlates with innovative uses of transparent glass by analyzing the many central Italian reliquaries from fourteenth-century Italy that feature small, round, glass windows, which functioned like monstrances and allowed one to view relics directly. Through such examples as Naddo Ceccarelli's *Reliquary Tabernacle with Virgin and Child*, this chapter connects the corporeal process of vision to its spiritual counterpart by investigating the role of sight in relic worship.

Chapter 6 examines a group of Franciscan reliquaries from the fourteenth century that combine the use of transparent glass and gilded glass in a single work and finds a complex fusion of ideas and techniques from Chapters 3 and 5. These reliquaries, which generally conform to a specific type, feature a single sheet of glass with portions of verre églomisé and transparent glass. Because of the dual use of glass serving as both a window and a mirror, these devotional tools could simultaneously resonate with different optical theories in order to offer their viewers complex devotional tools. To gain a more complete understanding of the reception of these works, this chapter situates them within the debate on visual theory and the Franciscan context by noting the order's many members who studied and wrote on the subjects of light and sight.

The concluding chapter summarizes the primary themes that run throughout the book. It argues that trecento Italy witnessed a cultural and artistic intersection between medieval and Renaissance aesthetics and, as such, its art featured aspects of each. The reliquaries discussed in this chapter combine the glittering reflective surfaces found in earlier examples of gilded glass with panels of transparent glass that display relic fragments to the viewer and, in doing so, draw from ideas presented in several of the preceding chapters. In addition to responding to the various artistic sources and evidencing new combinations unique to their cultural contexts, they also resonated with shifting optical theories, changing notions of the eye and vision, theological interpretations, and technological innovations. By examining the various ways artists, theorists, and viewers interacted with and conceptualized glass at this time, this concluding chapter explores how the various uses of and associations with glass were mutually informing each other. It further demonstrates that an understanding of these relationships provides new and important insights into the larger phenomenon of trecento visuality, or socially conditioned ways of seeing the world.

Ultimately *Seeing Renaissance Glass* argues that the theoretical and spiritual conceptions of vision and the visual arts were inextricably linked to developments in optical science, the practical experience of vision, and the medium of glass.

Brief History of Glass

Throughout the book the historical precedents of the type of glass under discussion are referenced, but it may be helpful to give a brief overview here as well. Though it is not clear how the invention of glass occurred, it is thought that glass was first made around 3000 BCE in Mesopotamia. Medieval and Renaissance perceptions of the medium's invention were influenced by Isidore of Seville's *Etymologies*, which describes how travelling merchants accidentally discovered how to make glass while camping on a sandy bank of the Belus River in Syria. Unable to find rocks to support their kettles, the merchants used natron from the ship's cargo. When they started the fire, the combination of natron, fire, and the pure sands of this region produced the first instance of glass.[17]

Some of the earliest extant Egyptian glass is opaque and was made using either the core-formed technique or by using a mold. Typical examples of the types of glass object include glass beads, bottles, other small containers, mosaic glass, and glass inlays for jewelry or furniture. As would be the case with glass throughout much of its history, the shapes, colors, and functions of the earliest glass objects suggest that these products were made in imitation of other materials, primarily precious stones or minerals. In fact, when Isidore of Seville describes the different colors of glass, he compares them to sapphires, onyx, and other gemstones.

Ancient Greek glass production, though slow at first, peaked during the Classical and Hellenistic periods. Not many glass vessels have been recovered from Mycenaean Greece, and the few objects that have been found suggest that at this time Greece did not have its own glass-production centers but rather imported premade glass from Egypt. Most of the objects, which consist of pendant-like ornaments and small figures dating from ca. 1300 BCE, were made from a bright translucent blue glass that matched the chemical composition of contemporaneous Egyptian glass.[18]

Glass production in Classical and Hellenistic Greece grew and emulated the Egyptian model in terms of technique and products until about the third century BCE. Burial finds suggest that early Greek glass objects were brightly colored vials with small openings used for the storage of perfume made using the core-formed technique and meant to imitate more precious materials.[19] By the mid-third century BCE, a more characteristically Greek glass industry had developed. Craftsmen developed new techniques such as network glass and mosaic glass, the latter of which was often used to make plates or bowls by fusing together many multicolored canes to form a complex and intricate spiral pattern.[20] It was also at this time that clear glass and gold sandwich glass developed, a topic treated in more detail in Chapter 3.[21]

It was in imperial Rome where glass production, of both colored and transparent glass, reached its zenith.[22] Glass was used in a variety of different ways in ancient Rome; it was fashioned into expensive luxury items like jewelry and lavish furniture inlays and used for household utilitarian storage containers, windows, and mirrors.[23] The Romans could create a vast range of objects because they had many different methods of working with glass available to them. The most revolutionary technique used in the production of ancient Roman glass was the blown-glass technique, which was developed near Syria in the first century BCE.[24] The Roman glass industry flourished as craftsmen refined the techniques of blowing, painting, engraving, gilding, and casting glass and, in so doing, provided long-lasting inspiration for the re-emergence of the glass industry in the Levant and the Latin West.[25]

The Byzantine and Islamic glass-making centers in the East preserved the Roman knowledge base and cultivated specifically eastern innovations in style and technique. Their fine luxury products, particularly lamps and so-called perfume jars, became highly valued commodities on the trade circuits, but these glass objects could also take on special religious significance, as seen in the case of pilgrim flasks. As discussed in Chapter 4, one of the main ways these glass bottles entered the Latin West was as reliquaries carried home from the Holy Land by religious pilgrims. The importation and inspiration of such objects, along with the import of cullet—pieces of broken glass that could be melted down and refashioned into new objects—as well as the arrival of glassworkers from the East, dramatically re-energized the production of glass in the West.

After the fall of Rome, glass production in the Latin West slowed considerably, but it did not die out entirely. While it is true that very few examples of medieval glass exist, Alan Macfarlane and Gerry Martin contend that the lack of archeological evidence supporting medieval glass manufacture is due to changes in medieval burial customs and glassmaking practices.[26] Oftentimes glass objects were interred with their owners and thus protected from centuries of handling. Tombs were effective time capsules, protecting fragile glass objects from destruction. However, with the rise of Christianity, burial customs changed. Christians did not regularly bury objects along with their deceased, effectively eliminating one of the main methods for preserving glass objects.[27] Also contributing to the decreased survival rate was the fact that, during the medieval period, glassworkers began melting and reusing glass objects to a greater degree. Furthermore, craftsmen began making their glass with potash from woodland plants instead of from sea plants, as was common practice in the Mediterranean region. The potash obtained locally from the woodland plants was more readily available, but the glass it produced was more prone to decay.

Venice was home to one of the most robust glassmaking centers in medieval Europe, an important point for a study of Italian glass. The Venetian glass tradition reached its peak in terms of skill and market value during the later Renaissance period, but, due to its connections with the Levant, it was already flourishing by the thirteenth century. It was also during the thirteenth century that we find some of the earliest Italian churches incorporating stained glass, with one of the most famous examples at San Francesco in Assisi. The next chapter will look more closely at the phenomenon of stained glass in an Italian context.

Notes

1. Vincent Ilardi, *Renaissance Vision from Spectacles to Telescopes* (Philadelphia: American Philosophical Society, 2007), 38.
2. Samuel Y. Edgerton, *The Mirror, the Window, and the Telescope: How Renaissance Linear Perspective Changed Our Vision of the Universe* (Ithaca, NY: Cornell University Press, 2009), 6.
3. Alessandro Bagnoli, *La Maestà di Simone* Martini (Milan: Silvana Editoriale, 1999).
4. It should be noted that Kessler does consider other reflective materials as substitute materials for glass mirrors. See Herbert Kessler, "Speculum," in *Speculum* 86 (2011): 10.
5. David C. Lindberg, *Theories of Vision from Al-Kindi to Kepler* (Chicago: University of Chicago Press, 1976), 87–90.
6. Ibid., 91–102.
7. Ibid., 92.
8. Ibid., 104–5.
9. David C. Lindberg notes, "In place of these discredited theories, [Albert] attempts to establish the Aristotelian doctrine that vision is caused by an alteration (*immutatio*) of the transparent medium by the visible object and the propagation of this alteration to the watery substance of the eye." See ibid., 105.
10. Lindberg, *Theories of Vision*, 109, 112. For more on Roger Bacon, see John Henry Bridges, *The "Opus Majus" of Roger Bacon* (Cambridge: Cambridge University Press, 1964); David C. Lindberg, *Roger Bacon's Philosophy of Nature* (Oxford: Oxford University Press, 1983); and David C. Lindberg, *Roger Bacon and the Origins of Perspectiva in the Middle Ages* (Oxford: Oxford University Press, 1996). For the most influential theorists preceding Bacon—that is (in chronological order), Seneca, Pliny the Elder, Chalcidius's translation of Plato's *Timaeus*, Augustine, William of Conches, Robert Grosseteste, and Albertus Magnus—see Lindberg, *Theories of Vision*, 87–107. For more on the most significant optical treatises of the fourteenth century, see ibid., 122–46. Albertus achieved a similar feat just before Bacon but did not fully integrate Alhazen.
11. Ibid., 109.
12. Ibid., 111–12.
13. Lindberg, *Theories of Vision*, 161.
14. Alan Macfarlane and Gerry Martin note, "He looked at various curved surfaces and the principles of refraction and reflection from these surfaces, using concave and convex mirrors. He looked at clear mirror images to see how the image is reflected in the mirror. Mirrors, prisms and lenses allowed the new mathematics and geometry to develop." See Alan Macfarlane and Gerry Martin, *Glass: A World History* (Chicago: University of Chicago Press, 2002), 42.
15. Robert Belle Burke, *The "Opus Majus" of Roger Bacon*, vol. 2 (Whitefish, MT: Kessinger Publishing, 2002), 546–57.

16. Ibid., 553–57. In one such example, Bacon described that when "the eye is at the center of a concave mirror, it is the only thing visible to itself; for no form is reflected to the center except that which comes from the center. The perpendicular, in fact, returns upon itself." Ibid., 555.
17. Stephen A. Barney and others, eds., *The Etymologies of Isidore of Seville* (Cambridge: Cambridge University Press, 2010), 328.
18. Hugh Tait, ed., *Five Thousand Years of Glass* (London: British Museum Press, 1991), 24–25.
19. Jennifer Price, "Glass," in *A Handbook of Roman Art*, ed. Martin Henig (London and New York: Phaidon Press, 2006), 205.
20. Tait, *Glass,* 48.
21. Ibid., 49. For more details on the other types of developments at this time see ibid., 50–61.
22. As Jennifer Price notes, during Roman times glass "exerted a greater influence on daily life than at any other period before the Renaissance." See Price, "Glass," 205. As Alan Macfarlane notes, "Roman glass technology was in many ways unrivalled until the nineteenth century." See Macfarlane and Martin, *Glass*, 13.
23. Macfarlane and Martin, *Glass*, 15–16.
24. Macfarlane and Martin, *Glass*, 13; Price, "Glass," 205–7.
25. The influence of Roman gold glass on later medieval glass is discussed further in Chapter 3.
26. Macfarlane and Martin, *Glass*, 19–21.
27. For exceptions describing some glass vessels from Christian burial sites, see David Whitehouse, *Medieval Glass for Popes, Princes, and Peasants*, with contributions by William Gudenrath and Karl Hans Wedepohl (Corning, NY: Corning Museum of Glass, 2010), 56.

References

Bagnoli, Alessandro. *La Maestà di Simone* Martini. Milan: Silvana Editoriale, 1999.

Barney, Stephen A., W. J. Lewis, J. A. Beach, Oliver Berghof, trans. *The Etymologies of Isidore of Seville.* Cambridge: Cambridge University Press, 2010.

Bridges, John Henry, ed. *The "Opus Majus" of Roger Bacon.* Cambridge: Cambridge University Press, 1964.

Burke, Robert Belle. *The "Opus Majus" of Roger Bacon.* Vol. 2. Whitefish, MT: Kessinger Publishing, 2002.

Edgerton, Samuel Y. *The Mirror, the Window, and the Telescope: How Renaissance Linear Perspective Changed Our Vision of the Universe.* Ithaca, NY: Cornell University Press, 2009.

Ilardi, Vincent. *Renaissance Vision from Spectacles to Telescopes.* Philadelphia: American Philosophical Society, 2007.

Kessler, Herbert. "Speculum." *Speculum* 86 (2011): 1–41.

Lindberg, David C. *Theories of Vision from Al-Kindi to Kepler.* Chicago: University of Chicago Press, 1976.

———. *Roger Bacon's Philosophy of Nature.* Oxford: Oxford University Press, 1983.

———. *Roger Bacon and the Origins of Perspectiva in the Middle Ages.* Oxford: Oxford University Press, 1996.

Macfarlane, Alan, and Gerry Martin. *Glass: A World History.* Chicago: University of Chicago Press, 2002.

Price, Jennifer. "Glass." In *A Handbook of Roman Art,* edited by Martin Henig, 205–19. London and New York: Phaidon Press, 2006.

Tait, Hugh, ed. *Five Thousand Years of Glass.* London: British Museum Press, 1991.

Whitehouse, David. *Medieval Glass for Popes, Princes, and Peasants.* Corning, NY: Corning Museum of Glass, 2010.

2

Stained Glass

Duccio, Simone Martini, and Taddeo Gaddi

Although more famous examples of stained glass windows are found in cities such as Chartres and Paris, artists in thirteenth- and fourteenth-century Italy also made impressive use of this medium. And perhaps surprisingly, the artists responsible for the designs of the stained glass windows in Siena's Cathedral, the Lower Church at Assisi, and Santa Croce were none other than the same individuals responsible for the naturalistic paintings at these locations: Duccio, Simone Martini, and Taddeo Gaddi.[1]

Italian stained glass has traditionally been marginalized in the scholarship due to a variety of factors, not the least of which is the prized position of fresco painting. The extensive frescoes at San Francesco in Assisi and the Scrovegni Chapel in Padua, for instance, reflect the traditional hallmarks of naturalism so frequently associated with Renaissance style. In the Upper Church of San Francesco, a single register of scenes depicting events from the life of Saint Francis runs the length of the nave showing a sequence of narrative images, each of which is a vignette wherein the main characters seem to perform their lines in front of a sparse, yet three-dimensional backdrop that references the actual architecture of medieval Assisi. In Padua Giotto and his workshop filled the walls of the Arena chapel with three registers of scenes. The top register depicts stories from the life of Saints Anne and Joachim, the middle row shows events from Christ's early life, and the lowest portion of the wall—that closest to the viewer—records Christ's Passion and Resurrection. Like the frescoes in Assisi, Giotto's compositions feature naturalistic figures and suggestions of realistic, albeit rudimentary, space, setting, and perspective. Furthermore, Giotto painted his outdoor scenes with blue skies, as opposed to the gold skies that were the norm in earlier, more abstract Italo-Byzantine paintings.

In general the paintings in Assisi and Padua reflect a growing cultural interest in humanism with their naturalistically modeled objects and three-dimensional space. In many ways these fresco programs set the stage for later, High Renaissance masterpieces such as Leonardo's *Last Supper*, Michelangelo's *Sistine Ceiling*, and Raphael's *School of Athens*. However, while the development of the naturalistic style of art is certainly central to an understanding of Renaissance art and culture, too often artworks that don't align with this trajectory are overshadowed. This is in no small part due to Giorgio Vasari's bias for the revival of the naturalistic style and the prominence of the so-called major art forms of painting, sculpture, and architecture, which relegated Italian stained glass to the periphery despite the prominent locations and artists involved with these works. As Nancy Thompson discusses in her important work on stained glass in Renaissance Florence, Vasari's preferences are clearly reflected in his limited discussion of windows and in his privileging of stained glass windows that feature qualities traditionally found in paintings. As Thompson puts it, Vasari appreciated stained glass windows that "masked [their] very materials."[2]

Another factor contributing to the marginalized place of trecento stained glass is the tendency to focus on the cult of the artist and imagine—somewhat incorrectly—a single, solitary author imbuing the work with evidence of their particular artistic genius through discernable brushstrokes and signature passages. In the case of stained glass windows—and most large-scale projects of the period—the endeavor was a collaborative effort, not a sole enterprise. After artists designed their compositions, glaziers contributed to the evolution of the finished products as they translated the compositions into glass panels. The relationship between the glazier and the designer was something of a partnership.[3] Thus as Thompson argues, in the creation of stained glass windows, the hand of the artist was not always immediately apparent, especially prior to the fifteenth century.[4]

The argument that follows here looks to reposition early Renaissance stained glass windows using examples such as Duccio's window at the Cathedral of Siena, Simone Martini's windows in the Chapel of Saint Martin in the Lower Church of San Francesco in Assisi, and Taddeo Gaddi's designs in the Baroncelli Chapel at Santa Croce in Florence in order to illustrate how the medium of stained glass was adapted to an Italian context and to investigate how it contributed to the evolution of Italian art. Formal analyses suggest that these artists approached stained glass windows differently than did their northern counterparts, creating an intentional dialogue between a windows' light and the same artist's paintings at the same site.[5] Such an examination not only reveals significant connections between the windows and their larger artistic context but also between the glass of the windows and other types of glass.

Brief History of Colorful Glass

Before turning to a detailed study of the windows, however, a brief history of the medium is necessary in order to establish the artistic context of the works. Glass was used to glaze windows as early as ancient Roman times, although ancient window glass was comparatively thicker than today's glass, manufactured in smaller pieces, and slightly bluish in color.[6] Documentary accounts attest to the use of glass in specifically Christian contexts by the sixth century.[7] Gregory of Tours (538–594) records how a glass window of a church was broken during a robbery, and Venantius Fortunatus (530–ca. 609) describes the beauty of a stained glass window in Paris.[8] Physical evidence, that is extant fragments of figurative stained glass, survives from as early as the first decade of the twelfth century in the form of depictions of Old Testament prophets from the windows of Augsburg.[9]

One of the most famous patrons associated with stained glass windows is Abbot Suger, who during the twelfth-century renovations of Saint Denis in Paris incorporated a program of windows and wrote about the process of commissioning them and their symbolism. The innovative Gothic-style pointed arch used in the church did not require extensive amounts of solid wall to support it; instead the structural load was carried by the expressive ribs, and solid surfaces were replaced with stained glass windows, the lighting effects of which were both beautiful and symbolic.

Suger's interpretation of this new aesthetic treatment was inspired by what he thought to be the writings of Saint Denis, a third-century Christian martyr and the namesake of the church. In reality the author was probably a Syrian theologian of the early-sixth century who was conflated with Dionysius the Areopagite mentioned in the Acts of the Apostles. Pseudo-Dionysius, as this author is now called, explained light's symbolic potential from a mystical and Neoplatonic point of view. Heavily influenced by these ideas, Suger commissioned artworks with glorious colors and dramatic lighting effects in order to transpose the viewer's thoughts from the mundane realm to the divine truths found in the heavenly sphere. In writing about his renovations of the church at Saint Denis, Suger noted how

> the loveliness of the many-colored gems has called me away from external cares, and worthy meditation has induced me to reflect, transferring that which is material to that which is immaterial, on the diversity of the sacred virtues: then it seems to me that I see myself dwelling, as it were, in some strange region of the universe which neither exists entirely in the slime of the earth nor entirely in the purity of Heaven; and that, by the grace of God, I can be transported from this inferior to that higher world in an anagogical manner.[10]

For Suger, then, colorful gems and stained glass windows were more than simply beautiful; they were transcendental and instrumental to cultivating devotion.[11] Furthermore, the windows of Saint Denis were not only capable of inspiring meditation on the divine, they were also divinely sanctioned, as Suger attested:

> For the most liberal Lord Who, among other greater things, has also provided the makers of the marvelous windows, a rich supply of sapphire glass, and ready funds of about seven hundred pounds or more will not suffer that there be a lack of means for the completion of the work.[12]

The fact that Suger referred to the blue glass as sapphire reflects an important connection between glass and gems, which may help explain the high value of glass.[13]

The writing of Saint Bonaventure (1221–1274), whose theology of light was highly influential within the Franciscan order, reveals insight into how late medieval Franciscans may have interpreted stained glass windows. And, as Thompson effectively argues, it was also the inspiration behind the commissioning of vast programs of stained glass in San Francesco, Assisi and Santa Croce, Florence, as well as in other programs.[14] As Bonaventure noted, "And just as one sees that the light of the sun, when penetrating a stained glass window, is broken into diverse colours, so does the Divine Ray shine in individual material creatures in many ways and in many properties."[15] As Thompson explains, there is a sequencing inherent in this comment. The first stage of one's anagogical ascent was appropriately the first thing one would see upon entering the physical space: the windows' colorful light, which symbolized the divine as manifested within the realm of the natural world. After using one's perception of the light as a starting point, their thoughts could ascend to more elevated matter.[16] More will be said about Bonaventure's theology of light in Chapter 3; what follows here is more concerned with the properties of *colored* light, specifically.

In addition to late medieval analogies such as that described by Bonaventure, colored glass, primarily due to its associations with precious gems, was believed to have had a long history of miraculous cures and spiritual benefits.[17] The view that gems and precious materials held virtuous qualities had existed at least since the time of the ancient Greeks when Theophrastus, writing in the fourth century BCE, described the properties of stones and gems in *On Stones* (*De lapidibus*).[18] In one instance particularly relevant to the optical interests of this study, he noted that *smaragdos*, or emerald, is "good for the eyes, and for this reason people carry seals made of it, so as to see better."[19]

Pliny the Elder's *Natural History* transmitted Theophrastus's ideas to the medieval West. In his text Pliny detailed the many functions of minerals known in ancient Rome. Stones offered a variety of remedies; they offered protection from harm by warding off evil, or when ground up and ingested, they could alleviate pain from injuries such as scorpion bites.[20] Like Pliny, Dioscorides, writing in the first century CE, also outlined the medicinal properties of stones in his *Materia medica*.[21] And in a similar vein, another famous ancient Greco-Roman source, the *Magic Papyri*, conveyed the great power of stones and the steps required for their activation.[22]

The apotropaic qualities of gems and jewels were passed down from the ancients to medieval Christians such as Marbode (1035–1122) and Albertus Magnus (ca. 1206–1280). Albertus's lapidary was particularly comprehensive in its treatment of minerals and notable for his explanation of their properties in relation to the four humors.[23] Although Albertus was careful to point out that stones were not living entities, he did emphasize that they were still very powerful on a variety of different levels. Some minerals could protect one from thieves while another type could bring the thief good fortune while conducting a robbery. Some stones functioned as a form of birth control while others ensured a healthy pregnancy. Still other types of stones could be used as a truth serum, an aphrodisiac, or a sedative.[24] Cameos made from onyx were said to promote dreams when put under one's pillow.[25]

The Bible added yet another layer of positive meaning to gems, glass, and precious minerals. Revelations 21:9–21 describes the Heavenly Jerusalem with many references to precious metals, gems, and glass.

> It shone with the glory of God, and its brilliance was like that of a very precious jewel, like a jasper, clear as crystal. … The wall was made of jasper, and the city of pure gold, as pure as glass. The foundations of the city walls were decorated with every kind of precious stone. The first foundation was jasper, the second sapphire, the third agate, the fourth emerald, the fifth onyx, the sixth ruby, the seventh chrysolite, the eighth beryl, the ninth topaz, the tenth turquoise, the eleventh jacinth, and the twelfth amethyst. The twelve gates were twelve pearls, each gate made of a single pearl. The great street of the city was of gold, as pure as transparent glass.

Evident in this description is the fact that gems, gold, and glass were worthy of a place in heaven and therefore should be highly valued on earth. In fact, gemstones were even equated with saints and the Heavenly Jerusalem, the walls of which are described as gems.[26] The powerful associations with gemstones certainly would have imbued artworks featuring them—or imitations of them using glass—with great value and prestige.

Based on the fact that lapidaries of the thirteenth century relate the powers of stones to their colors, Spike Bucklow extends the powerful associations of the gems to colored-glass imitations of gems.[27] Thus, red glass could have served as surrogate for ruby, green glass for emeralds, and so on.[28] This is important to note because, along with stained glass, another popular art form, enamel, also used glass in a way that emulated gems.

The enameling technique had a long history, with extant enamel works dating as early as the second millennium BCE. Several variations on this technique were available during the medieval period, the main two of which were cloisonné and champlevé. In both these enameling techniques, colored glass was cast, pulverized into powder, poured into compartments, and fired. After firing, the melted glass adhered to the artwork, and its grains fused to form a glossy, richly colored, jewellike texture.[29]

Instructions for making cloisonné were included in the medieval treatise *On Divers Arts* attributed to the twelfth-century German Benedictine monk, Theophilus.[30] It required one to create a series of small compartments with walls made from small strips of a precious metal, usually gold, to fill each cavity with the appropriate color of glass powder, and to fire the work to solidify the vitreous powder.[31] Cloisonné had been perfected by Byzantine artists and the technique—and objects made with it—arrived in the Latin west by at least the fifth century.

Champlevé, on the other hand, was primarily produced at and has since become associated with two main stylistic schools based in the Limoges and Mosan regions. In both the Limoges and Mosan styles, the pulverized glass was poured into cavities that had been carved into the metal's surface. Because the compartments were indented, the metal on which the artist worked needed to be very thick. As gold was very expensive, champlevé was usually made using gilded brass or bronze rather than gold. The cavities were also more shallow resulting in the use of less enamel.

The *Stavelot Triptych*, now in The Morgan Library and Museum, New York, serves as a good case study of the two main types of enamel techniques because both are featured within this object, and it demonstrates one of the main ways in which these enamel techniques were used: to decorate holy reliquaries. The two small triptych reliquaries affixed to the central panel, which feature cloisonné, are Byzantine imports and therefore nicely mirror the journey of the technique, that is, migrating from the East to the Latin West.[32] When these small holy triptychs arrived in the West, they were deemed worthy of a larger, more lavish, display context and were therefore set within a larger bejeweled triptych. Typical of this time, the holy objects were decorated with precious materials and narrative artworks, in this case, six narrative roundels which use the champlevé method according to the Mosan style.

Thus, when it came to enshrining some of the church's most prized treasures, the artists employed gems along with colored glass. With this artistic tradition in mind, along with the host of other aforementioned symbolism, it is not hard to see that the stained glass windows were imbued with a sense of sacredness.

Duccio's Window for the Cathedral of Siena

Duccio's stained glass window depicting the *Assumption of the Virgin* along with other scenes (Figure 2.1) for the choir of the cathedral of Siena from ca. 1287–1288 is an early example of a stained glass window designed by an Italian painter, and analysis of it within its larger artistic context begins to reveal how Italian artists assimilated Gothic stained glass traditions but modified them to a more painterly approach.[33] Unlike more traditional Gothic rose windows, Duccio's does not evoke the shape of a flower with outspread petals. Instead, the circular space is divided into registers, each with three large frames in a manner similar to the walls of Giotto's *Arena Chapel* or the panels in Duccio's own later *Maestà* altarpiece. The narratives in the center include, from the bottom to the top, the Dormition, the Assumption, and the Coronation. Seated evangelists with their characteristic symbols occupy the four corners. Moving clockwise starting in the upper left they are Saint John with the eagle, Saint Matthew with the angel, Saint Mark with the lion, and Saint Luke with the ox. The four patron saints of Siena occupy the middle register. To the left of the Assumption are Saints Bartholomew and Ansanus and to the right of the Virgin are Saints Crescentius and Savinus.

Throughout the window the figures occupy the majority of the visual field, and the lead tracery primarily serves as contour lines. Duccio's figures are proportionally large, he limits potentially distracting patterning, complex details are kept to a minimum, and simple blue backgrounds silhouette the figures, making them even more legible. In other words, in this window, the medium of stained glass has been adapted to the motives of the painter. Nowhere is this more apparent

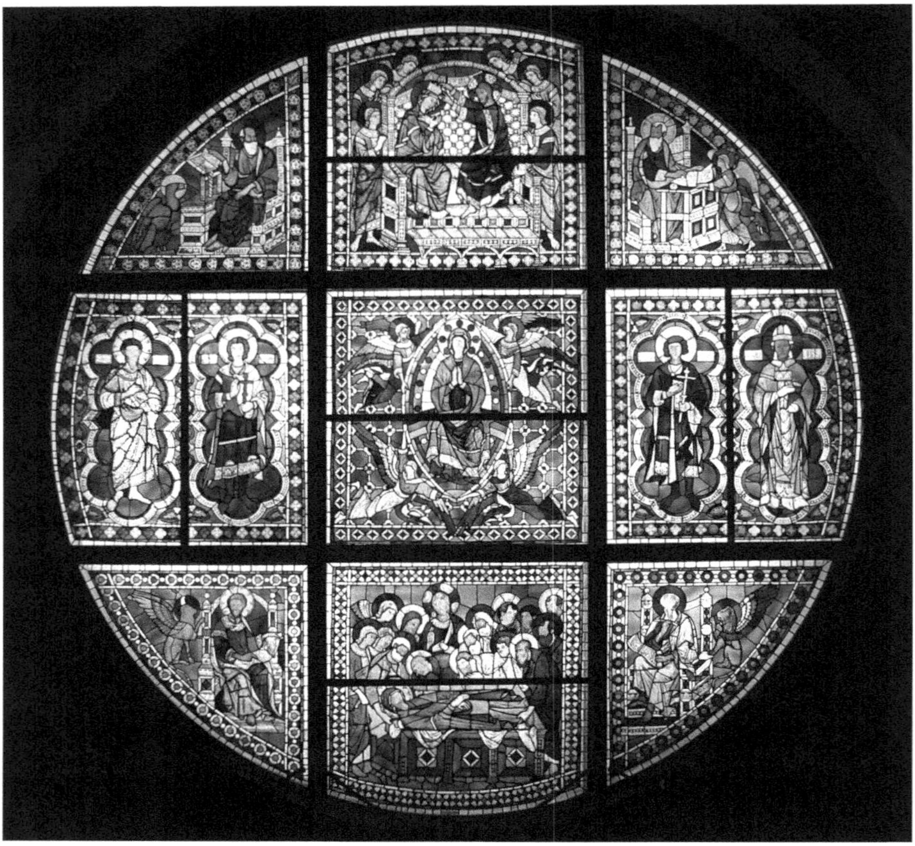

Figure 2.1: Duccio, *Rose Window with Scenes of the Dormition, Assumption, and Coronation of the Virgin*, ca. 1288, Cathedral, Siena. Source: José Luiz Bernardes Ribeiro via Wikimedia Commons (CC BY-SA 4.0).

than in the ways Duccio's imagery challenges the plane of the glass, for instance in the way the front legs of the Madonna's throne occlude the decorative border meant to contain them. This tension between the imagery and the border is also evident at the top of the scene where the angels' halos extend past the red and yellow pattern as well as in other passages throughout the window. The figural compositions seem to emerge from their decorative framework and assert themselves almost like a perspectival painting, and yet the medieval symbolism of light and color so characteristic of this medium was still a very potent artistic device.[34]

The symbolic associations with light were a hallmark of the medieval period, stemming as so much theology did from Neoplatonic influence. Not only was light a powerful symbol for God, but through its ability to illuminate the surrounding area and reveal the rich colors of the altarpiece, light was closely connected with the beauty of color and the presence of the divine. Duccio's window for the cathedral of Siena capitalizes on this rich tradition of light and color symbolism and at the same time diverges from Gothic traditions with its more pictorial treatment of the window's composition, which gives the window a sense of agency—the ability to relate to the viewers in new, dynamic ways.

The projection of the window's presence into the surrounding space was further enhanced by Duccio's altarpiece for this same location. Less than twenty years after the installation of his window, Duccio completed his *Maestà*, a massive polyptych altarpiece for the high altar, which stood just below the round window (Figure 2.2).[35] When designing this altarpiece, the artist would have certainly realized the potential for a conversation between the works. First, visitors to the cathedral would have had a continuous view of the entire nave, enabling them to simultaneously encounter both works as there was no evidence of a choir screen prior to 1367.[36] Second, other works by Duccio suggest that he looked for ways to connect his artworks with elements outside their frames.

The *Maestà* altarpiece is double sided, with each side catered to a specific type of viewer. The large scene of the Madonna Enthroned would have been visible to the congregation situated in the nave while the narrative panels on the reverse were smaller and more detailed, suitable to a more select and privileged viewing space, such as the choir where the more learned clergy had access to get closer and

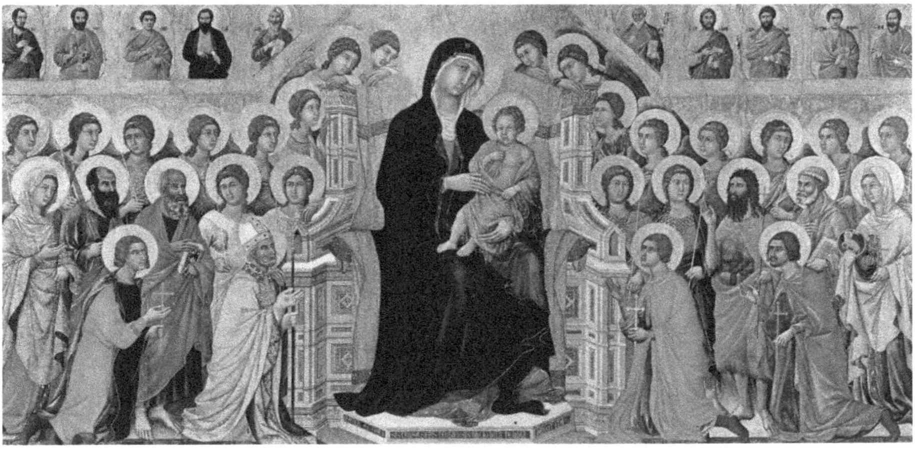

Figure 2.2: Duccio, *Maestà*, Detail of central panel showing Madonna, Child, and Saints, ca. 1308–1311, Cathedral, Siena. Source: Public Domain via Wikimedia Commons.

study the images. Although the arrangement of the narrative panels on the back of the *Maestà* is debated, it is fairly certain the first scene, *Entry into Jerusalem,* was located in the lower left hand corner. Considered in this context, the dominant compositional lines, which run at a severe diagonal from the lower left to the upper right, effectively create a directional force that mirrors the viewer's movement through the rest of the piece, which likely culminated at the upper right corner. To make this relationship clear, Duccio reinforces the visual entry of the viewer and their intended progression by including an open door prominently in the foreground of the scene.

A second example of Duccio's desire to bridge the painted space with the viewer's reality is found in the small panel now in the Metropolitan Museum of Art depicting a Madonna and Child set behind an illusionistic parapet, which simultaneously unifies and divides the viewer and the holy pair. The Madonna and Child are clearly situated behind the barrier and thus separate from the viewer, but because the ledge seems so familiar and naturalistic, it creates the sense that this scene could actually take place in the real world. Duccio's approach in the *Entry into Jerusalem* and the small devotional panel suggest that he, like other artists of the trecento, may have considered how his artworks would relate to the viewer in an experiential, almost physiological manner. It is therefore important to consider this artistic strategy when considering his stained glass window and altarpiece.

When viewed together in this context, Duccio's *Maestà* altarpiece and his stained glass window for the Sienese cathedral create a unified program that reinforces the prominent Marian imagery found elsewhere in the church. On bright days the colored light from the window could have flowed into the church, illuminating the equally rich colors of the polyptych's rear face and reflecting in all directions off the gilded portions of the frame. The scenes depicting the life of Christ in the altarpiece would have been bathed in what may have appeared as supernatural light, suggesting that the symbolic divine light revealed spiritual truth.

Adding to this kaleidoscopic medley of colors, the vertical axis created by the window and altarpiece could have resonated with a Neoplatonic interpretation. The painting, situated on ground level, presents the viewer with a scene based on earthly associations, while the window, situated high overhead, features greater focus on things of an ethereal nature. The altarpiece depicts a solid, weighty Madonna rendered with naturalistic modeling and set within a defined sense of space. She is surrounded by figures that stand in overlapping rows, as if a crowd of people has gathered around her large, sturdy, marble throne. The city's four patron saints kneel like humble subjects to either side and thus serve as models of behavior for the viewer.

In contrast, the window focuses on the divine nature of the Virgin's spirit and its eternal afterlife, a theme highlighted by the narrative moments selected. In the stained glass, the Virgin is no longer presented as a human mother. Likewise, her role as intercessor is de-emphasized; she is a heavenly figure to be venerated. The prominent blue backgrounds and repeating geometric patterns convey the sense that these images float in the sky above almost like apparitions or visions rather than naturalistic occurrences.

Another form of dialogue—which is central to the thesis of this book—existed between Duccio's window and his altarpiece: both incorporated or referenced the medium of glass. Though Duccio's painting did not feature actual panels of glass, it features several illustrations of glass. The depiction of the throne (Figure 2.2) on which the Madonna sits is decorated with gilded panels of glass arranged in intricate patterns, known as Cosmati. Inspiration for such a throne would not have been hard to find. As will be discussed in more detail in Chapter 3, many examples of Cosmati work were found throughout Italy, especially in Roman tombs and ciboria. But perhaps more pertinent to a study of Duccio's work in the cathedral of Siena is the fact that mere steps from the cathedral's high altar stood Nicola Pisano's marble pulpit, which as will be discussed in Chapter 3, features some of the earliest Italian decorative gilded glass decorating the spandrels just below the much more famous sculptural reliefs.

In addition to the depiction of gold glass inlays in the Virgin's throne, the back of Duccio's *Maestà* features three scenes with table settings—*The Last Supper* (Figure 2.3), *Christ Appearing to the Apostles*, and *The Wedding at Cana*—each of which show drinking vessels made of transparent glass.[37] The apostles and wedding guests drink red wine from glass cups, likely in imitation of actual cultural practice. The so-called *Aldrevandin Beaker* in the British Museum is a Venetian product that features a similar size, shape, and slightly tapered contour line as glasses depicted by Duccio. The production of such glasses was a relatively recent development in glassmaking technology, which will be discussed in greater detail in Chapter 5.[38]

Though these connections on their own do not affect a monumental shift in our conception of Duccio's artistic production, they do become more informative when situated alongside the increasing use of glass by artists discussed throughout this book. Taken in its entirety, the thesis of the book could indeed shift scholarly attention from individual artistic styles and the social context to an integrated approach to the period's art and technology and religion and theory.

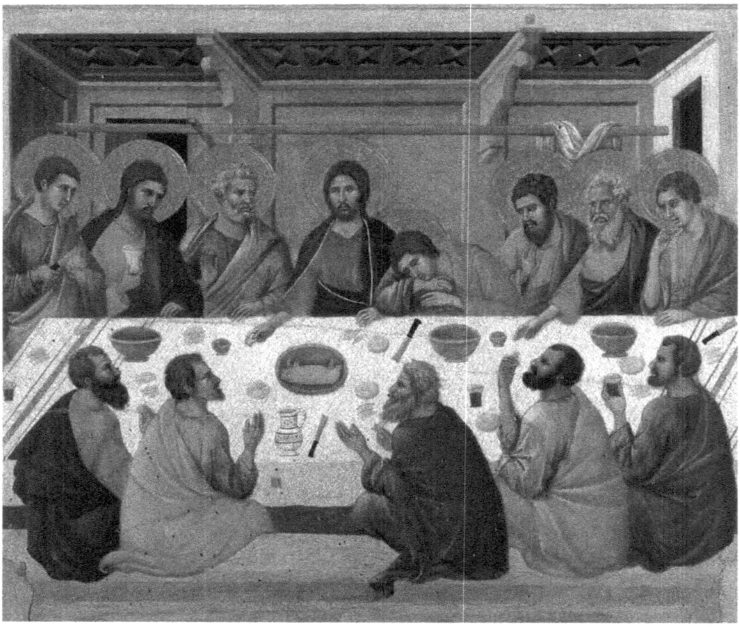

Figure 2.3: Duccio, *Maestà*, Detail with *Last Supper*, ca. 1308–1311, Cathedral, Siena. Source: Public Domain via Wikimedia Commons.

Simone Martini's Windows in the Chapel of Saint Martin

As with Duccio's window and altarpiece in Siena's Cathedral, the stained glass windows designed by Simone Martini for his chapel dedicated to Saint Martin in the Lower Church of San Francesco in Assisi (Figure 2.4) illustrate another case of a famous Italian painter designing stained glass windows in conversation with his surrounding commissions. And also like Duccio's window, which Simone certainly saw, Simone's composition strives for clarity of form and composition within the stained glass medium.[39] Simone's color choice for the windows almost declares his desire for legibility within the stained glass medium; the alternating red and blue backgrounds and yellow frames contribute to a sense of simplicity and organization.

Because of the more intimate relationship between Simone's windows and frescoes, it is possible to analyze them together as a coherent unit. Not only were both the windows and paintings designed by the same artist as part of a coherent program but they were also confined to a small chapel rather than spread across the interior of a large church. This enabled Simone to create a sense of unified space by integrating the light from the windows into his pictorial worlds in two

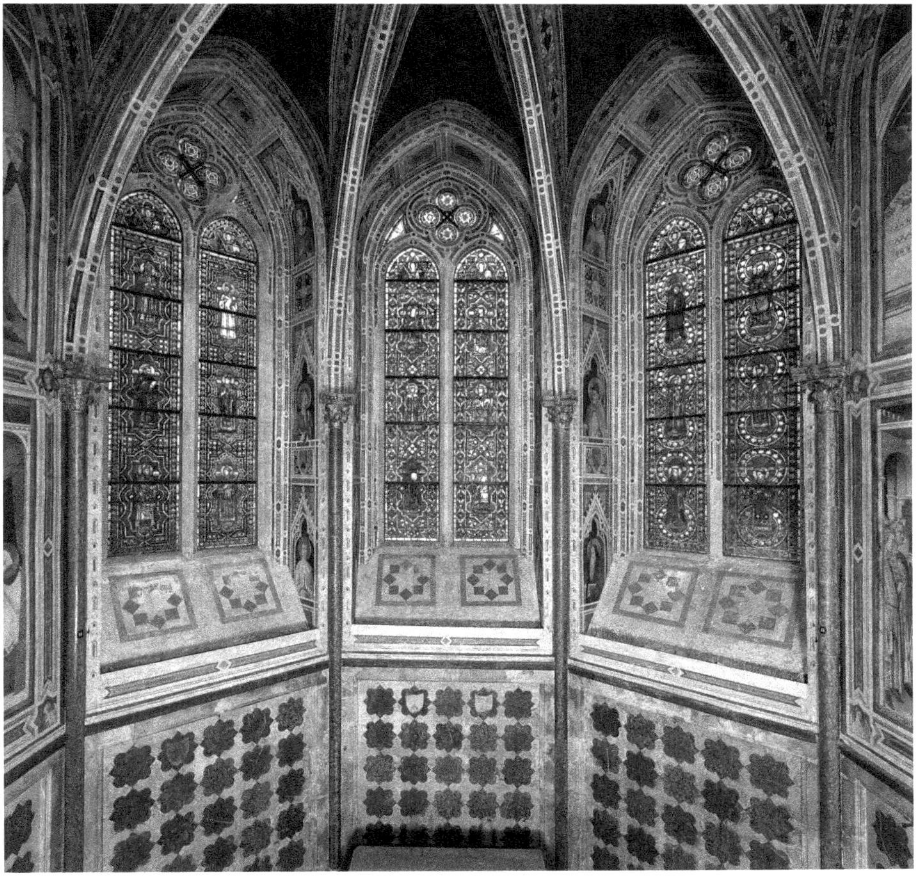

Figure 2.4: Simone Martini, Stained Glass Windows in *Saint Martin Chapel*, ca. 1317, Lower Church, San Francesco, Assisi. Source: Franco Cosimo Panini Editore © Management Fratelli Alinari.

ways. The first of these ways was by illuminating the frescoes to either side of the window as if they were lit by the window's light. The second way was by incorporating the symbolism of light into the narratives. The scenes to the right of the windows, namely, *Saint Martin and the Miraculous Mass* and *The Knighting of Saint Martin*, both feature Martin praying while he gestures up and to the left. In other words, in both scenes Martin prominently gazes in the direction of the stained glass window as he looks for divine inspiration.

The scenes to the left portray *Saint Martin in Meditation* on the top and *The Dream of Saint Martin* on the bottom and represent moments when Martin is about to receive divine insight. In the upper register Simone painted Martin enwrapped in a meditation so deep that his acolyte must physically interrupt it so the bishop can perform mass. In the lower register Christ, wearing the cloak that Martin had earlier given to a beggar, visited Martin in a dream and confirmed that Martin's charitable gesture had not gone unnoticed. These scenes feature Martin in an altered state ready to receive divine insight. By situating Martin towards the left of each composition, Simone again made the connection between the windows' light and the saint's divine inspiration explicit. In the scenes to the left, Simone showed Martin praying to God and asking for inspiration, while the frescoes to the right depicted him receiving divine wisdom. The program as a whole illustrates Saint Martin's deep connection to the divine through his ability to see beyond his painted reality to the colorful light of the windows and the symbolic presence of God they represent.

To make the connections between his painted realities and the stained glass unmistakable, Simone referenced the shape of the chapel's physical windows in his painted illustrations of windows, as seen in the lancet windows in the background of *Saint Martin in Meditation*. Another prominent use of the lancet shape is found on the underside of the entry arch leading into the chapel, where standing saints seem to almost converse with the viewers entering the chapel. These painted saints in the entry serve as pendants to the saints depicted in the windows; in each case both the painted and stained glass, individual saints stand within a decorative frame and gesture to each other or the visitor below.

However, an important difference may suggest there were two different conceptions behind the painted saints versus the stained glass ones. The saints in the windows do not look out at the viewer; their gazes are kept firmly within the confines of the window's space. It is as if their attitude and the glowing, supernatural light of the windows signaled that these figures were not as accessible as the painted saints in the archway. The physical location of the saints in the chapel reinforced this; the viewer would have had the opportunity to stand directly under the painted saints but the stained glass ones were set behind the altarpiece and out of reach. However, not all the figures depicted in the windows face inward. In fact, the angels at the top of the two lancets stare straight out into the chapel's space, but these militant angels do not engage with the viewers in order to offer comfort or their services as intercessors like the painted saints. Rather, these figures hold swords and shields indicating their role as guardians of the divine realm. Their bright yellow shields glow in the dark chapel, evoking the sense of glaring reflections bouncing off their polished metal. Thus, it should be noted that, for all the ways Simone forged a connection between the windows and the surrounding space, he also created the sense that the imagery in the windows occupied a different realm from the painted reality.

As in the case of Duccio, Simone's decorative complex at Assisi was part of a larger network of glass and one should contextualize Simone's stained glass windows within the artist's other uses of or references to glass objects. Prior to his work on the Chapel of Saint Martin, Simone had completed his fresco of the *Maestà* in the Palazzo Pubblico in Siena. As will be discussed in the next chapter and was previously mentioned in Chapter 1, Simone set twenty-five pieces of gilded glass into the frescoed wall to enhance the lighting effects of his painting (Figure 1.2). He also used a glass roundel to embellish his panel of Saint Louis of Toulouse, which was roughly contemporary with the chapel.

Simone would have also been witness to a particularly striking use of gilded glass each time he went to work in the Chapel of Saint Martin because set into the ceiling of the Lower Church were pieces of glass backed with silver that were meant to resemble stars in the night sky (Figure 2.5). Though Simone was not responsible for making or designing these panels of glass, it is possible that he took note of them because he reproduced the appearance of the Lower Church's ceiling in the depiction of *The Funeral of Saint Martin* and *Saint Martin and the Miraculous Mass.*

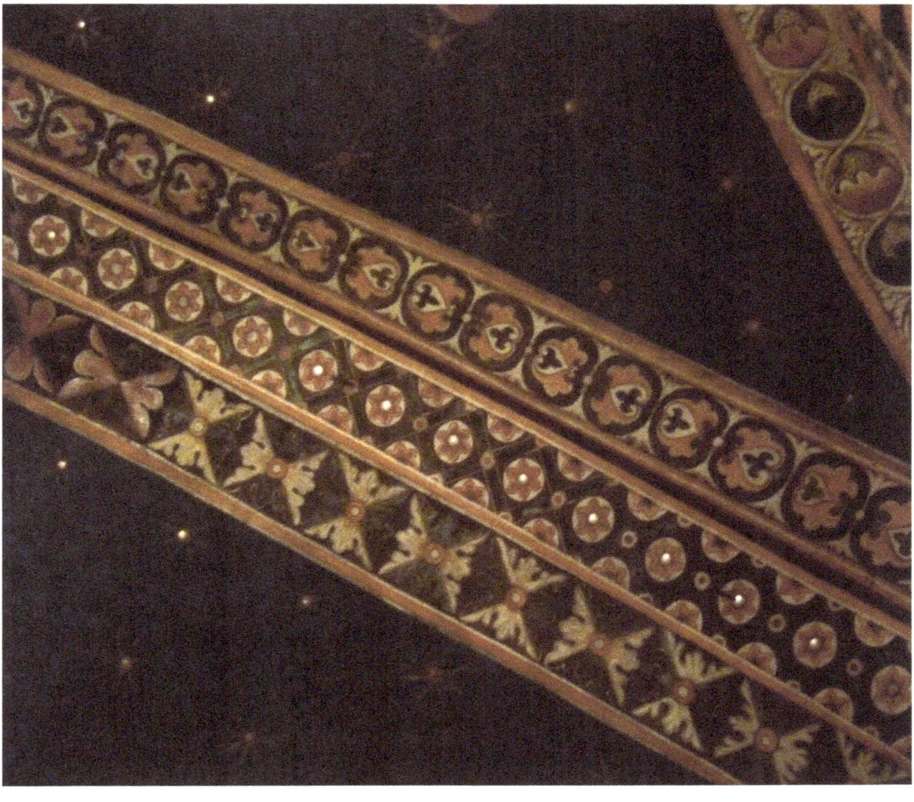

Figure 2.5: Ceiling of Lower Church with light reflecting off silver-backed glass insets, Lower Church, San Francesco, Assisi. Source: Author.

Taddeo Gaddi's Stained Glass in the Baroncelli Chapel

As demonstrated by the cases of Duccio and Simone, Italian stained glass was often integrated into the surrounding decorative program, dismantling the boundaries between media. Nowhere is this more fully realized than in Taddeo Gaddi's Baroncelli Chapel in Santa Croce of Florence (Figure 2.6). The Baroncelli family commissioned this burial chapel in honor of the Virgin Mary, and the decoration throughout aptly celebrates her. The frescoes on the walls depict various aspects of her life, while the painted altarpiece commemorates her Coronation, and the sculptures positioned at the entry portal reenact the Annunciation. As in the case of Duccio's stained glass window, Taddeo's window also shows suggestions of perspective and pictorial devices as well as a desire to extend its agency beyond the frame to interact with the surrounding environment.

Diana Norman and Paul Hills have described the many ways in which Taddeo's chapel carefully incorporated the stained glass window and its lighting into a cohesive program about the mystery of divine revelation.[40] As with the case of Saint Martin's chapel, lighting from the window has been used in both descriptive ways and symbolic ones throughout the frescoes. The scenes to the left of the window—which show the Annunciation to Mary on the top register, the Annunciation to the shepherds in the middle, and the Annunciation to the Magi on the lowest register—feature a visionary moment where an angelic visitor seems to almost emerge from the window to deliver a divine message. As already mentioned, Simone Martini employed a similar strategy in the chapel of Saint Martin, and even closer in time and space, Taddeo's master Giotto used the ambient light as inspiration for his frescoes in the Bardi and Peruzzi Chapels, which were located just across the transept from Taddeo's own chapel.[41] However, unlike in Giotto's chapel, Taddeo used the windows' light to illuminate and describe the illusionistic scene, and he also used the same light as a symbolic metaphor for the presence of the divine.[42]

Attesting to Taddeo's interest in unusual lighting effects is the fact that his frescoes—particularly his *Annunciation to the Shepherds*—are some of the earliest Renaissance paintings to depict dramatic night scenes and sensitivity to the shadows produced by bright light falling on objects such as the strap of the shepherd's canteen. Furthermore, the contour lines found throughout this depiction are somewhat blurred and the colors are muted, visual effects which are consistent with a night scene lit by a sharp, bright light source. This comparison is an interesting one because some scholars have asserted that Taddeo may have in fact observed such a scenario during the solar eclipse of 1339.[43]

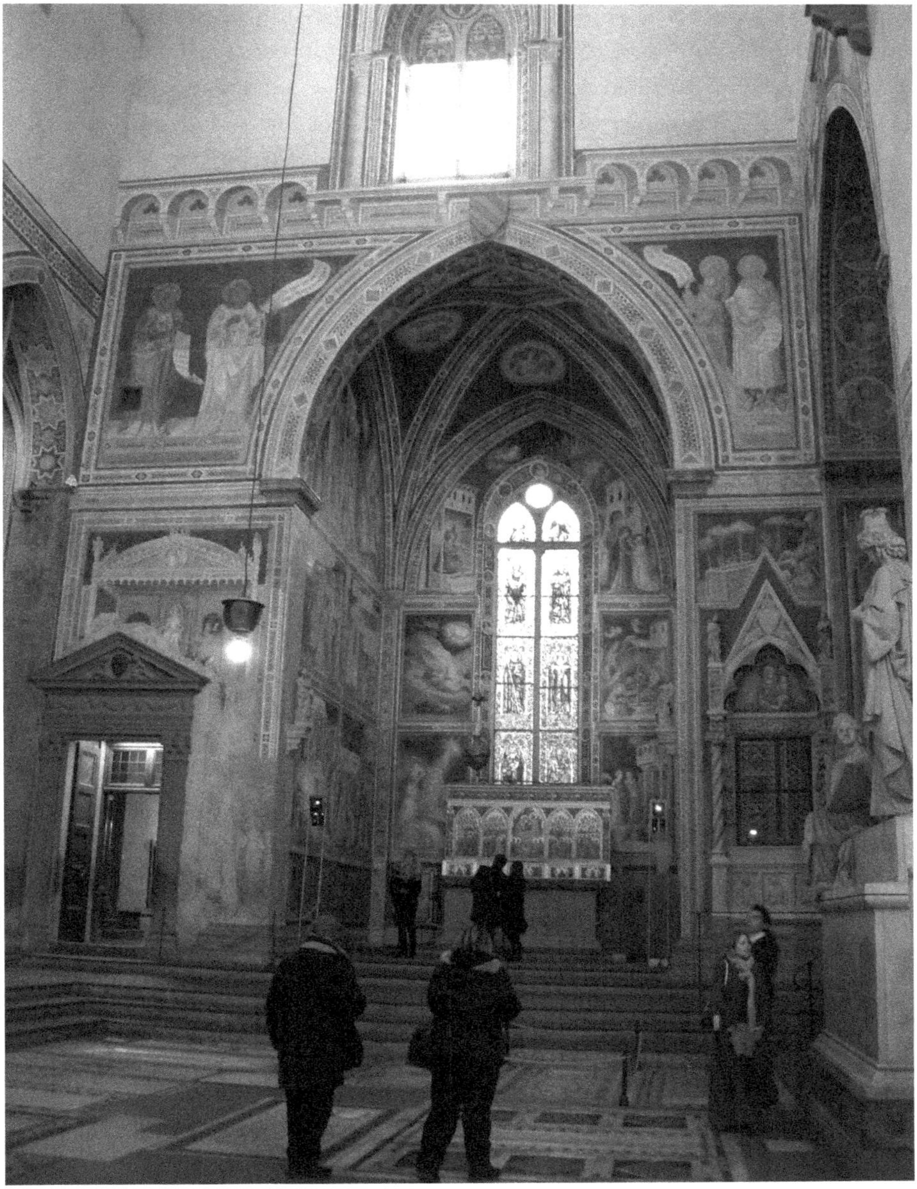

Figure 2.6: Taddeo Gaddi, *Baroncelli Chapel,* ca. 1330, Santa Croce, Florence. Source: Francesco Bini via Wikimedia Commons (CC BY-2.5).

Another aspect of Taddeo's chapel that attests to this artist's more in-depth interest in optics and light is the depictions of glass in the chapel's original altar-piece. In the now-detached pinnacle currently in the San Diego Museum of Art (Figure 2.7), God is depicted as an older, bearded, man surrounded by six angels. Four of the angelic beings shield their eyes from the divine glory of God, while two others hold circular objects that appear to be made of glass as if to reflect or somehow refract a direct view of God. Whether the circular objects are clear glass or reflective glass, this scene closely resonates with an illustration of the Beatific Vision as described by Saint Paul, where one can only view God "through a glass, darkly."[44] The angels experience the difficulty of viewing God's powerful illumination just as an artist would have experienced pain while trying to study the lighting of the sun.[45]

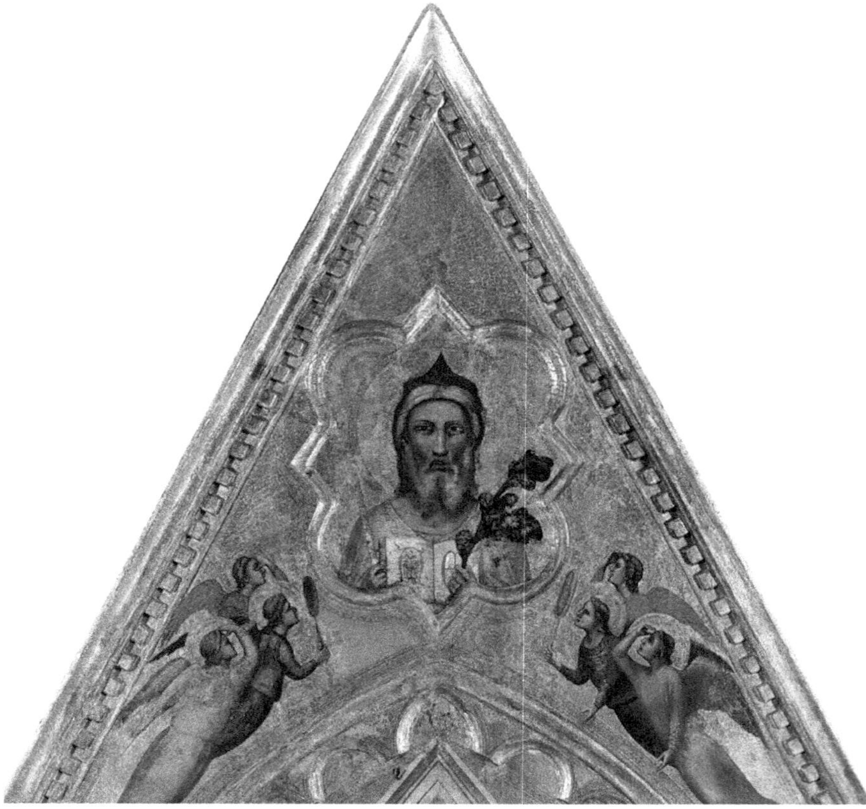

Figure 2.7: Giotto and Workshop, *Pinnacle for the Baroncelli Altarpiece*, ca. 1334, San Diego Museum of Art, San Diego, Gift of Anne R. and Amy Putnam (1945.26). Source: Public Domain via Wikimedia Commons.

Although the Baroncelli altarpiece has traditionally been attributed to Giotto, many scholars have noted the master was in high demand at this time, and the execution would have largely been carried out by his workshop. Taddeo Gaddi was taking his own commissions but he was still very involved with Giotto's shop, and therefore it is not hard to imagine that Taddeo played a large role in the altarpiece as well.[46] When this possibility is taken into consideration, the depictions in altarpiece pinnacle take on greater import and evince a larger network of glass.

Another way in which Taddeo's chapel engaged with the glass network was through the depiction of transparent glass in one of the illusionistic niches at the base of the wall. In the second niche, to the viewer's left upon entering the chapel, the fictive cupboards contain a liturgical vessel for wine made of transparent glass (Figure 2.8). By making the wine—the blood of Christ—directly visible through the clear glass, the artist allowed the viewer an opportunity to see the wine, which would soon be transformed into the blood of Christ, and meditate on the miraculous transubstantiation. Like the divine revelations occurring in the paintings on the walls above the fictive cupboards, the clear glass carafe offered the viewer a chance to come close to experiencing a physical presence of the divine. More will be said about how transparent glass facilitated relic worship in Chapters 4, 5, and 6. But here it is important to note that there were many transparent glass and crystal reliquaries with which Taddeo could have been familiar.

Conclusions on Trecento Stained Glass: A Network of Glass

To understand fourteenth-century Italian stained-glass windows in their full cultural context, it is necessary to consider the history of glass as well as the windows' aesthetic and symbolic potential alongside the various other uses of trecento glass. The Italian painters who were working with stained glass were also interested in other types of glass, either actual or illusionistic. Artists like Duccio, Simone Martini, and Taddeo Gaddi were all continuing in the long tradition of glass arts as well as adapting the medium to their unique situations.

In most Gothic stained glass windows, the imagery functioned as a sort of "illuminated painting," to borrow Anne Friedberg's term.[47] However, as Friedberg points out, the stained glass window was a nonperspectival picture plane.[48] Friedberg does not speak specifically about Italian trecento windows. Had she done so, she might have noted how, in the hands of Duccio, the medium of stained glass took on a greater sense of perspective, particularly in his renderings of the evangelists at their

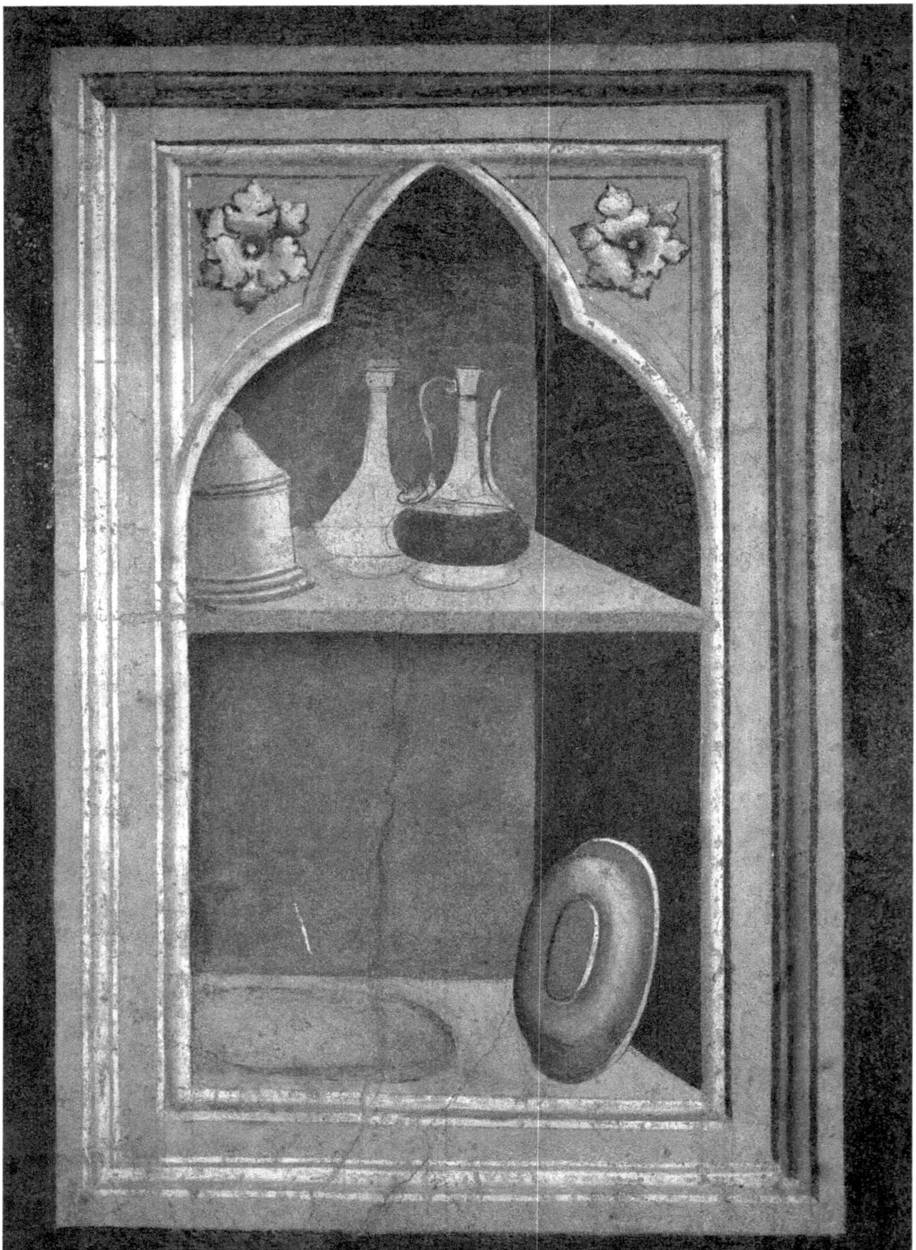

Figure 2.8: Taddeo Gaddi, *Illusionistic Niche with Liturgical Vessels*, ca. 1330, Baroncelli Chapel, Santa Croce, Florence. Source: Raffaello Bencini/Alinari Archives, Florence.

desks in the corners of the window for Siena's cathedral. When employed by Simone and Taddeo, the window also showed signs of perspective—for instance in the pedestals on which the saints stand. The windows in Assisi and Florence also became more three-dimensional in the sense that they interacted with their surrounding decorative complex. The windows' effects were not confined to the single plane of their glass panels; rather, they projected their colorful light, and their agency, into the pictorial reality around them and thus into the viewer's space.

This chapter has demonstrated that Italian artists well known for their illusionistic space, naturalistic modeling, and narrative skills applied these same objectives when working with glass. Stained glass windows provided them new opportunities to advance these interests by harnessing the powerful and beautiful symbol of colored light to their artistic goals. Many of the ideas presented in the subsequent chapters have their roots in the artworks, artists, and ideas presented here. As this study will suggest, when working with gilded glass and transparent glass, some of these same trecento artists assimilated these media to their already-existing objectives and sensibilities rather than simply adopting the traditional methods of working with glass. In these ways it is possible to see that glass was both a reflection of and inspiration for trecento artistic production and, in turn, visuality.

Notes

1. This chapter is not meant to be a comprehensive treatment of all Italian early Renaissance stained glass; many other works are not mentioned. Other projects include Maso di Banco's work in Santa Croce and the contributions of Agnolo Gaddi, Ghiberti, and Donatello at the Florence Cathedral. For more on Maso di Banco's work see John White, *Art and Architecture in Italy: 1250–1400* (New Haven, CT: Yale University Press, 1993), 420; for the projects in the Florence Cathedral, see Nancy Thompson, "Designers, Glaziers, and the Process of Making Stained Glass Windows in 14th- and 15th-Century Florence," *Journal of Glass Studies* 56 (2014), 237–51.
2. Thompson, "Making Stained Glass Windows," 238, 249–51.
3. For more on this relationship, see ibid., 238–39 and Reneé Burnam, "The Glazing of Siena Cathedral's *fenestra rotunda magna*: Preliminary Observations from a Production Standpoint," in *The Four Modes of Seeing: Approaches to Medieval Imagery in Honor of Madeline Harrison Caviness*, eds. Evelyn Staudinger Lane, Elizabeth Carson Pastan, and Ellen M. Shortell (Burlington, VT: Ashgate, 2009), 13–29.
4. Thompson, "Making Stained Glass Windows," 239, 246–47.
5. This is not to say that French stained glass windows do not work in tandem with other artworks in their immediate environment. As Virginia Chieffo Raguin points

out there were significant and inspiring connections between the imagery in stained glass and reliquaries, as in the case of the tomb of Thomas Becket and between the glass and the architecture and sculpture; see Virginia Chieffo Raguin, *Stained Glass: From Its Origins to the Present* (New York: Abrams, 2003), 17, 60. I do argue, however, that the way in which the Italian artists discussed here incorporated the windows and their lighting effects into the adjacent frescoes specifically is something distinct to this artistic environment.

6. Anne Friedberg, *The Virtual Window* (Cambridge, MA: MIT Press, 2006), 107. Paul Lacroix Jacob notes evidence of glass windows used in Pompeii, though not widespread use; see Paul Lacroix, *The Arts in the Middle Ages, and at the Period of the Renaissance* (London: Chapman and Hall, 1870), 252–53.
7. For early history of windows in Latin churches and the specific role of the Benedictines, see Alan Macfarlane and Gerry Martin, *Glass: A World History* (Chicago: University of Chicago Press, 2002), 20.
8. For several other early references to glass windows in churches see Sarah Brown, “Stained Glass,” in *The Oxford History of Western Art*, ed. Martin Kemp (New York: Oxford University Press, 2000), 108; and Jacob, *Arts in the Middle Ages*, 252–53, which also describes the accounts of Gregory of Tours and Venantius Fortunatus as well as other early examples of stained glass windows.
9. Raugin, *Stained Glass*, 59. It is thought that the depictions in these fragments were complimented by a series of apostles, Christ, and the Virgin Mary.
10. Erwin Panofsky, *Abbot Suger on the Abbey Church of St. Denis and Its Art Treasures* (Princeton, NJ: Princeton University Press, 1979), 63–65.
11. This idea was inspired by the writing of Pseudo-Dionysus who Suger mistakenly believed was Saint Denis. For more on this see Paul Rorem, *Pseudo-Dionysius: A Commentary on the Texts and an Introduction to Their Influence* (Oxford: Oxford University Press, 1993); and Paul Rorem, *Pseudo-Dionysius: The Complete Works* (New York: Paulist Press, 1987).
12. Panofsky, *Abbot Suger*, 53.
13. Raugin also cited the gem-stained glass connection using Suger’s sapphire comment; see Raugin, *Stained Glass*, 10, 14.
14. Nancy Thompson, “The Franciscans and Stained Glass in Tuscany and Umbria,” in *Mendicant Cultures in the Medieval and Early Modern World: Word, Deed, and Image*, eds. Sally J. Cornelison et al. (Turnhout: Brepols, 2016), esp. 24, 34–41. The scholarship of Nancy Thompson is indispensable in terms of understanding stained glass within a specifically Italian context. Also see Thompson, “Making Stained Glass Windows,” 237–51; and two forthcoming publications: Nancy Thompson, “The Creation of Stained Glass in Central Italy, 1250–1400,” in *Reading Medieval Sources: Stained Glass*, eds. Elizabeth Pastan and Brigitte Kurmann-Schwartz (Leiden: Brill, 2018); and Nancy Thompson, “Networks and Materials: Italian Stained-Glass Windows ca. 1280–1400,” in an edited volume of the papers presented at the Andrew

Ladis Trecento Conference in November, 2016, eds. Sarah Wilkins and Holly Flora (Turnhout: Brepols, 2018).

15. As quoted in Thompson, "Franciscans and Stained Glass," 31.
16. Ibid., 32–33.
17. See Martina Bagnoli, "The Stuff of Heaven: Materials and Craftsmanship in Medieval Reliquaries," in *Treasures of Heaven: Saints, Relics, and Devotion in Medieval Europe*, exh. cat., eds. Martina Bagnoli et al. (New Haven, CT: Yale University Press, 2010), 138.
18. Theophrastus, *On Stones,* trans. Earle Radcliffe Caley and John F. C. Richards, Graduate School Monographs, no. 1 (Columbus, OH: Ohio State University, 1956).
19. Ibid., 50.
20. Dimitris Plantzos, *Hellenistic Engraved Gems* (New York: Clarendon Press, 1997), 110.
21. Ibid.
22. Ibid., 32.
23. Spike Bucklow, "The Virtues of Imitation: Gems, Cameos, and Glass Imitations," in *The Westminster Retable: History, Technique, Conservation,* eds. Paul Binski and Ann Massing (Cambridge: University of Cambridge, 2009), 143.
24. On Albertus Magnus's summary of the properties of minerals see John M. Riddle and James A. Mulholland, "Albert on Stones and Minerals," in *Albertus Magnus and the Sciences: Commemorative Essays 1980* (Toronto: Pontifical Institute of Mediaeval Studies, 1980), 209.
25. Bucklow, "Virtues of Imitation," 143.
26. Bagnoli, "The Stuff of Heaven," 138–39.
27. Spike Bucklow extends the powerful associations of the gems to colored-glass imitation of gems. "Virtues of Imitation," 98.
28. Bucklow, Ibid., 143–44.
29. Other variations on these techniques include *basse-taille*, *plique-à-jour*, and émail *en ronde bosse*. For more on this and the history of this tradition, see Janetta Rebold Benton, *Materials, Methods, and Masterpieces of Medieval Art* (Oxford: Praeger, 2009), 149, 153–55.
30. For more on the attribution and the various extant editions of this work see Theophilus, *On Divers Arts,* trans. John G. Hawthorne and Cyril Stanley Smith (New York: Dover Publications, 1979), xv–xxxv.
31. Ibid., 126–27.
32. Joyce Brodsky, "The Stavelot Triptych: Notes on a Mosan Work," *Gesta* 11, no. 1 (1972): 31; William Voelkle, *The Stavelot Triptych: Mosan Art and the Legend of the True Cross* (New York: Oxford University Press, 1980), 20.
33. Diana Norman, *Painting in Late Medieval and Renaissance Siena: 1260–1555* (New Haven, CT: Yale University Press, 2003), 76. The windows at San Francesco in Assisi predate the window in Siena, but these were designed by German artists around

ca. 1250; see White, *Art and Architecture,* 184. For more on Duccio's window also see Marilena Caciorgna, Roberto Guerrini, and Mario Lorenzon, eds., *Oculus cordis. La vetrata di Duccio: Stile, iconografia, indagini tecniche, restauro* (Ospedaletto: Pacini, 2007).

34. For other ways in which Duccio's design challenged the traditional physical structure of the window glass and its supporting mechanisms see Burnam, "Preliminary Observations," 18–20.
35. The altarpiece was dismantled in the eighteenth century, and portions of it (the majority of which are in the Museo dell'Opera del Duomo, Siena) can now be found in museums around the world.
36. Norman, *Medieval and Renaissance Siena,* 38.
37. Giotto also depicted similar glass drinking vessels in his *Last Supper* panel now in Munich's Alte Pinakothek.
38. Other surviving examples of small, mostly clear glass drinking vessels dating from the thirteenth and fourteenth centuries are found in David Whitehouse, *Medieval Glass for Popes, Princes, and Peasants* (Corning, NY: Corning Museum of Glass, 2010).
39. White, *Art and Architecture,* 361.
40. Paul Hills, *The Light of Early Italian Painting* (New Haven, CT: Yale University Press, 1987), 75–86; Diana Norman, "Those Who Pay, Those Who Pray and Those Who Paint: Two Funerary Chapels," in *Siena, Florence, and Padua: Art, Society and Religion: 1280–1400,* ed. Diana Norman, vol. 2 (New Haven, CT: Yale University Press, 1995), 169–79.
41. Hills, *Early Italian Painting,* 75.
42. He could have been exposed to interesting lighting effects through his father's mosaic works, such as the *Coronation of the Virgin* located in the Duomo above the main entrance on the interior or, as will be discussed shortly, the eclipse of 1339.
43. There is a letter, thought by some to have been written by Taddeo, describing how he injured his eyes while trying to observe the solar eclipse. Paul Hills notes that Taddeo's inspiration could have come from simply observing night scenes and their corresponding lighting effects in his daily life; Hills, *Early Italian Painting,* 81. For an overall summary of the debate see Norman, "Two Funerary Chapels," 179 and R. J. M. Olson and J. M. Pasachoff, "Comets, Meteors, and Eclipses: Art and Science in Early Renaissance Italy," *Meteoritics and Planetary Science* 37 (2002): 1567–1471.
44. 1 Cor. 12:13. For more on the panel see Norman, "Two Funerary Chapels," 172; and Christine Sciacca, ed., *Florence at the Dawn of the Renaissance: Painting and Illumination 1300–1350* (Los Angeles: J. Paul Getty Museum, 2012), 110–13, fig. 23.
45. See note 36, above, for more on this.
46. White, *Art and Architecture,* 413.
47. Friedberg, *Virtual Window,* 107.
48. Ibid., 107–8.

References

Bagnoli, Martina. "The Stuff of Heaven: Materials and Craftsmanship in Medieval Reliquaries." In *Treasures of Heaven: Saints, Relics, and Devotion in Medieval Europe*, exh. cat., edited by Martina Bagnoli et al., 137–47. New Haven, CT: Yale University Press, 2010.

Benton, Janetta Rebold. *Materials, Methods, and Masterpieces of Medieval Art.* Oxford: Praeger, 2009.

Brodsky, Joyce. "The Stavelot Triptych: Notes on a Mosan Work." *Gesta* 11, no. 1 (1972): 19–33.

Brown, Sarah. "Stained Glass." In *The Oxford History of Western Art*, edited by Martin Kemp, 108–123. New York: Oxford University Press, 2000.

Bucklow, Spike. "The Virtues of Imitation: Gems, Cameos, and Glass Imitations." In *The Westminster Retable: History, Technique, Conservation,* edited by Paul Binski and Ann Massing, 143–151. Cambridge: University of Cambridge, 2009.

Burnam, Reneé. "The Glazing of Siena Cathedral's *fenestra rotunda magna*: Preliminary Observations from a Production Standpoint." In *The Four Modes of Seeing: Approaches to Medieval Imagery in Honor of Madeline Harrison Caviness*, edited by Evelyn Staudinger Lane, Elizabeth Carson Pastan, and Ellen M. Shortell, 13–29. Burlington, VT: Ashgate, 2009.

Caciorgna, Marilena, Roberto Guerrini, and Mario Lorenzon, eds. *Oculus cordis. La vetrata di Duccio: Stile, iconografia, indagini tecniche, restauro.* Ospedaletto: Pacini, 2007.

Friedberg, Anne. *The Virtual Window.* Cambridge, MA: MIT Press, 2006.

Hills, Paul. *The Light of Early Italian Painting.* New Haven, CT: Yale University Press, 1987.

Lacroix, Paul. *The Arts in the Middle Ages and at the Period of the Renaissance.* London: Chapman and Hall, 1870.

Macfarlane, Alan, and Gerry Martin. *Glass: A World History.* Chicago: University of Chicago Press, 2002.

Norman, Diana. *Painting in Late Medieval and Renaissance Siena: 1260–1555.* New Haven, CT: Yale University Press, 2003.

———. "Those Who Pay, Those Who Pray and Those Who Paint: Two Funerary Chapels." In *Siena, Florence, and Padua: Art, Society and Religion: 1280–1400,* edited by Diana Norman, vol. 2, 169–194. New Haven, CT: Yale University Press, 1995.

Olson, R. J. M., and J. M. Pasachoff. "Comets, Meteors, and Eclipses: Art and Science in Early Renaissance Italy." *Meteoritics and Planetary Science* 37 (2002): 1563–1578.

Panofsky, Erwin. *Abbot Suger on the Abbey Church of St. Denis and Its Art Treasures.* Princeton, NJ: Princeton University Press, 1979.

Plantzos, Dimitris. *Hellenistic Engraved Gems.* New York: Clarendon Press, 1997.

Raugin, Virginia Chieffo. *Stained Glass: From Its Origins to the Present.* New York: Abrams, 2003.

Riddle, John M., and James A. Mulholland, "Albert on Stones and Minerals." In *Albertus Magnus and the Sciences: Commemorative Essays 1980.* Toronto: Pontifical Institute of Mediaeval Studies, 1980.

Rorem, Paul. *Pseudo-Dionysius: A Commentary on the Texts and an Introduction to Their Influence.* Oxford: Oxford University Press, 1993.

———. *Pseudo-Dionysius: The Complete Works.* New York: Paulist Press, 1987.

Sciacca, Christine, ed. *Florence at the Dawn of the Renaissance: Painting and Illumination 1300–1350.* Los Angeles: J. Paul Getty Museum, 2012.

Theophrastus. *On Stones*. Translated by Earle Radcliffe Caley and John F. C. Richards. Graduate School Monographs, no. 1. Columbus, OH: Ohio State University, 1956.

Theophilus. *On Divers Arts.* Translated by John G. Hawthorne and Cyril Stanley Smith. New York: Dover Publications, 1979.

Thompson, Nancy. "Designers, Glaziers, and the Process of Making Stained Glass Windows in 14th- and 15th-Century Florence." *Journal of Glass Studies* 56 (2014): 237–51.

———. "The Franciscans and Stained Glass in Tuscany and Umbria." In *Mendicant Cultures in the Medieval and Early Modern World: Word, Deed, and Image*, edited by Sally J. Cornelison, Nirit Ben-Aryeh Debby, and Peter Howard, 22–44. Turnhout: Brepols, 2016.

———. "The Creation of Stained Glass in Central Italy, 1250–1400." In *Reading Medieval Sources: Stained Glass*, edited by Elizabeth Pastan and Brigitte Kurmann-Schwartz. Leiden: Brill, 2018.

———. "Networks and Materials: Italian Stained-Glass Windows ca. 1280–1400." Turnhout: Brepols, 2018.

Voelkle, William. *The Stavelot Triptych: Mosan Art and the Legend of the True Cross.* New York: Oxford University Press, 1980.

White, John. *Art and Architecture in Italy: 1250–1400.* New Haven, CT: Yale University Press, 1993.

Whitehouse, David. *Medieval Glass for Popes, Princes, and Peasants.* Corning, NY: Corning Museum of Glass, 2010.

3

Gilded Glass

Nicola Pisano, Simone Martini, Orcagna, and Paolo di Giovanni Fei

Clear glass panels backed with gold leaf are ubiquitous throughout late medieval and early Renaissance art.[1] Though most commonly associated with the mosaic technique, fourteenth-century Italian artists also used gilded glass in commissions ranging from sculpture to fresco. Some of the most famous artworks featuring gilded glass include Giotto's *Crucifix* for Santa Maria Novella, Orcagna's *Tabernacle* for Orsanmichele, Simone Martini's *Maestà* for Siena's Palazzo Pubblico, and Paolo di Giovanni Fei's *Madonna and Child* in New York's Metropolitan Museum, each of which will be discussed in this chapter.

Art-historical scholarship does not typically address the visual qualities or theoretical implications of gilded glass found in the aforementioned works.[2] Because these panels can produce sharp, glaring reflections that obscure aspects of the artworks, professional photographs of these monuments are often lighted in such a way as to minimize or eliminate the reflective properties of the glass panels. The intentional omission of these lighting effects in photographs is both a product of and a contributing factor to a large body of scholarship primarily focused on stylistic analysis and connoisseurship.

Recent studies, however, have been more sensitive to considerations of the ways in which these artworks were originally viewed, resulting in greater attention paid to original lighting contexts.[3] In fact, neuroscientists are becoming increasingly involved in the study of art history. Margaret Livingstone's major contribution to the field of visual perception research, *Vision and Art: The Biology of Seeing,* established a fundamental bridge between optics and art viewing. Of more specific interest to the present study is an experiment conducted by Claus-Christian Carbon that analyzed the impact of lighting on various types of paintings in order to better understand the conditions of medieval viewers, or as he calls it, the "perceptual past."[4] As these efforts attest, while it is certainly difficult—if not impossible—to reimagine or reconstruct the original circumstances of fourteenth-century devotional artworks, it is nevertheless an important endeavor if one wants to understand the period's reception of gilded glass and the intentions of the artists who were using it.

The most pertinent aspect of Carbon's study for the present discussion is the conclusion that perception of paintings with a gold leaf background was especially affected by lighting conditions.[5] The reflective surface of the gold leaf added an important layer of visual information. Gilded glass panels operated in a manner similar to paintings with gold leaf backgrounds, creating striking visual effects of their own. In an age before electricity, when the primary light sources derived from candles or sunlight, the relatively dark spaces these works occupied would have been pierced by the sharp reflections produced by the glass panels as seen in a photograph of Simone Martini's *Maestà* which captures the reflective properties, albeit at the expense of a focused representation of the artwork's finer details (Figure 3.1).

Careful consideration of these visual effects is important for two reasons. First, the noticeable lighting effects created by glass would have dramatically shaped one's visual experiences of artworks and, second, both gold and light were imbued with rich spiritual symbolism. Light was believed to have a corporeal form, but in its purest state, this form was invisible to one's physical sense of sight, only revealing itself through divine intervention.[6] Any earthly manifestation of light was worthy of close observation and reverence by an early modern viewer because it served as a means of accessing the divine.

Connections between divinity and light can be found throughout ancient and medieval philosophy, with associations referenced in ancient Egypt and Greece. The first New Testament figure to make this connection was Christ himself when he proclaimed in John 8:12, "I am the light of the world." During the medieval period, thinkers such as Augustine and Dionysius the Areopagite (Pseudo-Dionysius) continued the dialog, as mentioned in the previous chapter. By the late medieval and early Renaissance periods, the Franciscans had emerged as the group at the forefront of optical studies with great interest in the physical and symbolic properties of light.

Saint Bonaventure (1221–1274), one of the most famous medieval commentators on light, wrote extensively about the theoretical implications of light in

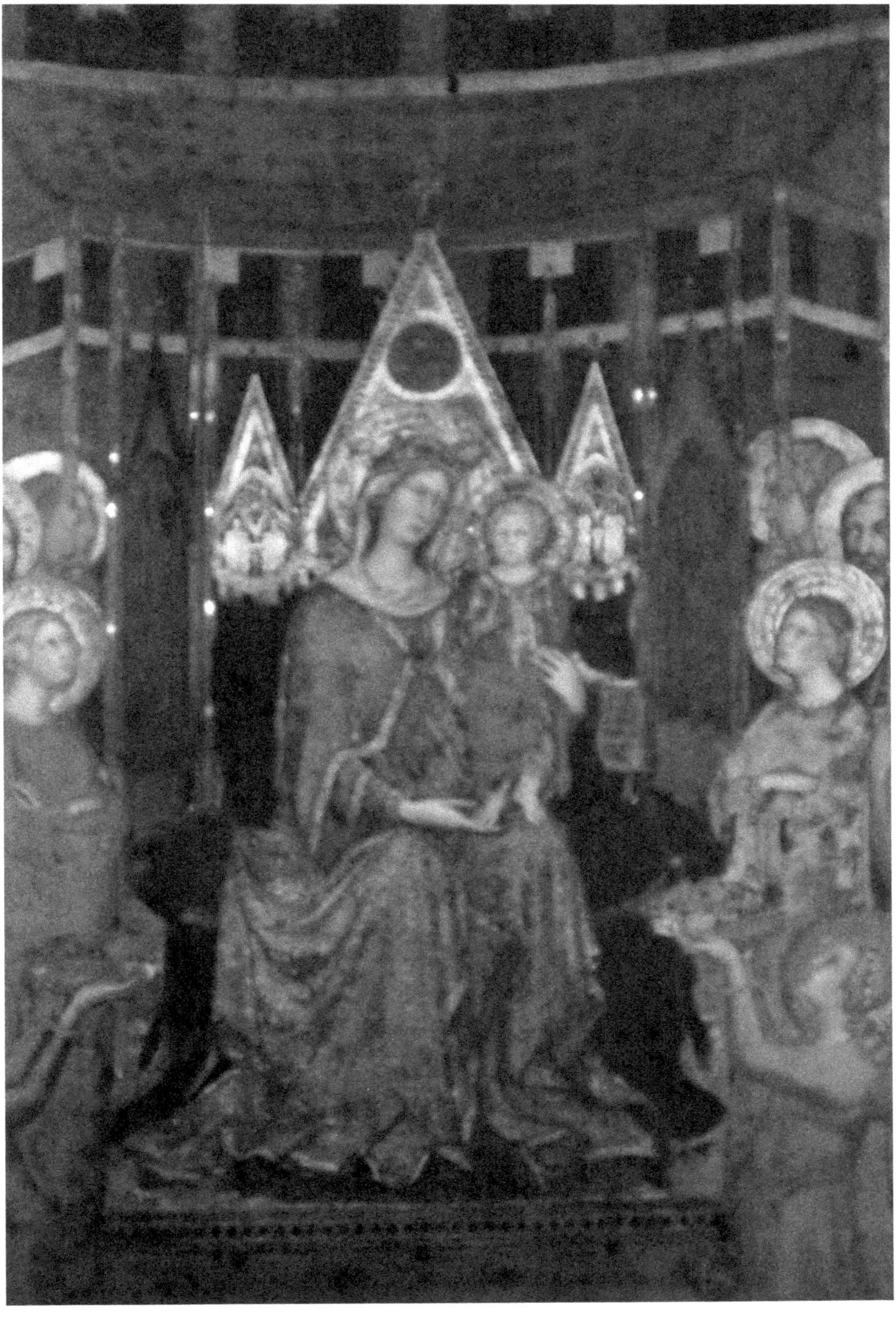

Figure 3.1: Simone Martini, Detail of *Maestà* showing diamond-shaped panels of glass reflecting light, ca. 1315, Palazzo Pubblico, Siena. Source: Author.

his commentary on the *Sentences* of Peter Lombard and made many references to the connections between light and the divine. He referred to God as the "Father of Lights"[7] in his prologue to *The Soul's Journey into God,* and in an attempt to explain the relationship between Christ and God, he noted in his *Tree of Life* that

> In this eternal kingdom, all good and perfect gifts come down in plenty and abundance from the Father of Lights (James 1:17) through Jesus Christ, who is the superessential Ray ... For he is a pure effusion of the brightness of the power of the omnipotent God.[8]

As this passage suggests, Bonaventure understood that there existed important difference between the various aspects of light, that is, between the source of light and its rays. God is the original light source, and Christ is likened to the emanating illumination. Following in the tradition started by Augustine, Bonaventure conceptualized the different aspects of light as lux, lumen, and color.[9] He distinguished them as follows:

> *Lux* can be considered as threefold, that is in itself, and in transparent media, and as terminated at the limits of the perspicuous: in the first mode it is *lux*, in the second *lumen*, in the third the hypostasis of *color*.[10]

In other words, lux is the essential nature of light, lumen is the radiance that stems from this source, and color is the effect produced when light falls onto solid objects.[11]

As the essential nature and source of light, lux was comparable to God in a spiritual analogy and, in a more practical corollary, to an object such as the sun. Or as in the case of John Pecham (c. 1230–1292), a thirteenth-century Franciscan natural scientist known for his work on optics, lux could also be equated to a candle flame.[12] Lumen, the emanation of lux, finds resonance with Christ, the ultimate emanation from God and the vehicle by which God's light entered the world. And, in a more empirical, observable sense, lumen corresponded to the rays of the sun or the aura radiating out from the candle flame.

According to Bonaventure, color was light resting on a solid object. Color, in this sense, pertained to the visual qualities of physical objects. As the natural world and all objects within it were thought to be a reflection of God, color was the visual manifestation of God's glory through his created world. One's experience of the natural world, and the way in which it reflected God's goodness, depended on the presence of light as color.

Using Giotto's *Crucifix* for Santa Maria Novella as a case study, one can imagine how a viewer might have reflected on these various aspects of light. Votive candles around the church could have evoked the notion of God, while their emanating light could have reminded one of Christ as God's emissary on earth. The colors of the painting could have evoked ideas about the beauty of the natural world, especially considering that they were used to create such a naturalistic rendering of the human body. Typical of his style, Giotto painted the roughly life-size crucified body of Christ in a highly realistic way, modeling the anatomy to convey a sense of three-dimensionality, weight, and gravity.

However, when it came to depicting the halo, Giotto perhaps realized that the naturalistic style was not well equipped to capture the essence of such a divine symbol. There was no earthly referent for the supernatural phenomenon of the halo and thus no degree of naturalism would suffice. Perhaps this conundrum inspired the artist to turn to the symbolic potential of gilded glass.[13] Giotto inserted panels of gold-backed glass into the wooden panel around Christ's head in a circular shape to convey the supernatural qualities of the halo.[14]

The sharp points of light reflecting off Christ's halo, however, do not fit neatly within any of the aforementioned categories outlined in Bonaventure's scheme of light metaphors. Another Franciscan writing in the late thirteenth century addressed this absence. Bartolomeo da Bologna (d. 1294) added a fourth type of light to Bonaventure's system: that of splendor.[15] Bartolomeo even makes specific reference to a gilded panel in his description of splendor, noting that

> when rays emanating from a luminous body reach another body that is smooth, polished and shining, such as a sword or gilded panel (*tabulam deauratam*), and rebound back from that body this is called *splendor*. And by such reflections on a polished and shining body the light (*lumen*) in space is multiplied and such multiplication of light is properly called *splendor*.[16]

Thus, while color is light falling on an opaque body, splendor occurs when light reflects off a luminous surface. The glittering effect one observes as splendor has a supernatural quality, as if it comes from some unknown source, making it a particularly suitable symbol for God's divinity.[17]

Medieval optical studies recognized the interdependence of light and sight; light made sight possible. Therefore, an analysis of one must be contextualized within an early modern conception of the other. Doing so reveals that medieval visual theories could have resonated in powerful ways with the lighting effects produced by gilded glass panels. From the fourth century to the end of the thirteenth century the dominant visual theory in the Latin West derived from Plato. His theory of vision, known as the extramission model, theorized that visual perception resulted from a process wherein the eye sent out visual beams from the eye, and then these "fiery rays" coalesced with external light and reached out to the object.[18]

Thus, when praying in front of Giotto's *Crucifix*, viewers could have conceived of the dramatic golden reflections emanating off the gilded glass panels as more than just indications of the holiness or special nature of the site. The sharp rays of light, or splendor, bouncing off the gilded glass could have been understood as visual manifestations of Plato's fiery rays emitted from the eye, hitting their object of perception, and returning to the viewer with information about the perceived object. In this context, the painting's gilded panels offered visual verification of the moment of contact between the viewer and the holy object, between the physical world they could see and the divine world beyond. As Photius declared in verse 5 of his Seventeenth Homily, from the year 867,

> For surely, having somehow through the outpouring and effluence of the optical rays touched and encompassed the object, it too sends the essence of the thing seen on to the mind, letting it be conveyed from there to the memory for the concentration of unfailing knowledge.[19]

Though Photius comes from an earlier, Byzantine tradition, his words represent a general understanding of the Platonic visual system, which was influential throughout the Latin West as well. The sparkling, glittering highlights could have function as a form of visual receipt, confirming the moment of visual contact.

As spiritually inspiring as these gilded pieces of glass may have been, it is important to note that they were not ideal in all contexts. Isidore of Seville discouraged their use in secular, academic settings, such as libraries, remarking how architects "would not think of putting gilt ceiling panels in libraries … because the glitter of gold wearies the eyes."[20] This comment simultaneously attests to the noticeable—one might even say arresting—effects of gilded glass panels and also suggests that such powerful visual displays were understood to be appropriate in certain settings, presumably ones that evoked a more abstract, metaphysical mindset rather than ones which required suitable lighting for close reading.

The Glittering Gold of Mosaics and Cosmati

Mosaics, Cosmati, and *verre églomisé* are important precedents for an examination of gilded glass in late medieval and early Renaissance Italian art because each of these techniques incorporates small clear glass panels backed with a sheet of gold leaf and would have been part of the visual vocabulary of artists such as Giotto,

Nicola Pisano, and Simone Martini. In the mosaic and Cosmati techniques, the piece of glass was entirely gilded, whereas in verre églomisé, or the reverse painted glass technique, only portions of the glass were treated with gold. Thus, while there are strong formal similarities among all three artistic techniques, important distinctions exist.[21] What follows provides a brief history of each technique and an examination of how Nicola Pisano, Simone Martini, and Paolo di Giovanni Fei adapted gilded glass to their artistic objectives.

Gilded glass tesserae were used extensively in early Christian, Byzantine, Islamic, and medieval European mosaics. Artists from a wide range of times, locations, and even religions, recognized how effectively clear glass panels backed with gold leaf could symbolize aspects of the divine. As early as Constantine the Great, the apses of Christian churches were covered with gilded tesserae, which may have functioned like symbolic messengers of divine enlightenment as they "appeared to extend the *photismos* imparted by Christ through baptism to the members of his church," as Erkinger Schwarzenberg puts it.[22] Eve Borsook's study of the ways in which artists manipulated the tesserae—that is by tilting them, affixing them with their reverse side showing, and using silver in combination with gold—reveals that their irregular placement was specifically designed to "enhance the glitter."[23] Borsook finds further evidence that gilded tesserae were intentionally used to convey divine symbolism through their reflective surfaces in the many mosaic *tituli* that accompany them, which proclaimed both the spiritual and formal qualities of the mosaics they adorned.[24]

Mosaics with gilded tesserae also had practical, more mundane benefits. The amplifying effects of the reflective tesserae would have illuminated the dark, mostly enclosed, interior spaces they decorated, providing greater visibility. And as Borsook explains, from at least the fourteenth century onwards, their role as a durable form of decoration was increasingly valued. Simultaneously, there was also growing interest in their costly nature and therefore a subsequent association with luxury and status.[25]

This is not to say that material value was not of interest during the earlier medieval period. After all, in the Byzantine tradition, gold was considered the most precious material in terms of both intrinsic and spiritual value.[26] Even tesserae without gold were highly valued because they were either made of some other precious material or they were fashioned using a high level of skilled craftsmanship. In fact, tesserae were so prized that they were treated as sought-after spoils. In one instance Charlemagne received tesserae which had been removed from churches in Ravenna, and in another, Venice was the beneficiary of a "shipload" worth of the precious pieces after the 1204 conquest of Constantinople.[27]

Interestingly, the shift away from an interest in divine symbolism to more mundane motivations observed in mosaic imagery coincides with the increasing application of individual gold-backed glass panels in Italian sculptures and paintings. Thus, it is worth considering that the use of gilded glass in works such as the pulpit, tomb, and paintings under discussion here was both a continuation of the mosaic tradition and a new application of the medium, one which carried on gilded glass's association with divine illumination, and one that suited the unique visuality of duecento and trecento Italy.

Although not as ubiquitous or well studied as mosaics, the cosmatesque technique—a variation of the ancient art of opus sectile, or cut work—was also a rich potential source of inspiration for artists using gilded glass, especially those artists who had spent time in Rome. Cosmatesque, or Cosmati work, derived its name from the Cosmati family of Roman artists. The technique involved assembling small pieces of stone, glass, or other materials into elaborate geometric patterns. Unlike mosaics, however, Cosmati typically does not feature figurative imagery or involve setting the tesserae at irregular angles; rather, the pieces are set flat against a solid surface, and the designs they create are dominated by repeating geometric shapes and abstract patterns. Also unlike mosaic, the pieces of gilded glass used in Cosmati are more often standard-size tiles with smooth edges, flat faces, and consistent, geometric shapes, not irregular shapes or surface textures.

This labor-intensive technique was most commonly used to decorate floors, tabernacles, altars, and tombs from the eleventh through the thirteenth century. Some of the most famous monuments decorated with Cosmati include the tomb of Hadrian V from ca. 1276 by Arnolfo di Cambio at San Francesco in Viterbo; the same artist's ciborium at San Paolo fuori le Mura from ca. 1285; the tomb of Honorius IV from the 1280s in Santa Maria in Aracoeli; the tomb of Clement IV from 1271–74 at San Francesco in Viterbo; the high altar of the Upper Church in San Francesco, Assisi, from 1253; and the floor of Westminster Abbey.[28] As this list indicates, although the technique is most commonly associated with Rome, monuments featuring Cosmati with gilded glass can be found throughout Italy and even farther abroad.

In addition to actual, physical, artworks with Cosmati, depictions of Cosmati can also attest to the medium's captivating qualities. Two of the most famous panel paintings to feature Cosmati include works by Duccio and Giotto. As mentioned in Chapter 2, Duccio's *Maestà* altarpiece for the cathedral of Siena shows the Virgin's throne covered in elaborate Cosmati with gilded glass as does the throne of Saint Peter in Giotto's *Stefaneschi Altarpiece*, along with many other depictions of furniture in trecento art.

Orcagna's *Tabernacle* in Orsanmichele, Florence, is another example of the long-lasting influence of the Cosmati technique. Though it now houses a painting of the Virgin Mary by Bernardo Daddi, the monument was associated with a lineage of paintings depicting the Madonna that were credited with miraculous healing powers. To imbue the marble structure with a sense of holiness worthy of these miraculous images, Orcagna's Cosmati included extensive use of glass panels throughout the spiral columns (Figure 3.2), in the reliefs, and especially in the depiction of the *Assumption of the Virgin* on the back of the tabernacle where reflective panels fill the sky and frame the scene. The style of Cosmati closely reflects the medieval Roman examples, which Orcagna may have seen firsthand on a 1350 trip to Rome to celebrate the Jubilee. As this monument was made midway through the fourteenth century, it will be discussed in more detail later in this chapter after an examination of the works by Nicola Pisano and Simone Martini, which predate it by several decades.

The Revival of Verre Églomisé

The technique of verre églomisé is similar to mosaic and Cosmati artworks in that gilded glass played an integral role in the composition; however, unlike the gold glass of the two aforementioned techniques, in verre églomisé the gilded panel was etched with imagery, sometimes floral motifs but in other cases figural compositions and even narrative scenes. To protect and seal the gold leaf imagery, the glass panel was backed with pigment. Thus, whereas the glass panels used in the two other techniques receive a consistent application of gold leaf, in verre églomisé gilding does not necessarily cover the entire panel of glass.

An early iteration of this technique dates back to ancient Greece, when Hellenistic glassmakers developed the method known as sandwich glass. In this process, rather than protecting gold leaf decoration with pigment, an artist sandwiched gold leaf designs between two sheets of glass. For example, in ancient Greek luxury works such as the *Sandwich Gold-Glass Bowl* from the third century BCE in the British Museum (inventory no. 1871,0518.2), elaborate foliage patterns of gold are sandwiched between two pieces of mostly transparent glass.[29]

Glass was more ubiquitous and affordable in Roman times, in large part due to the development of glass blowing, a factor that may have contributed to greater variety and experimentation.[30] Roman artists frequently depicted more

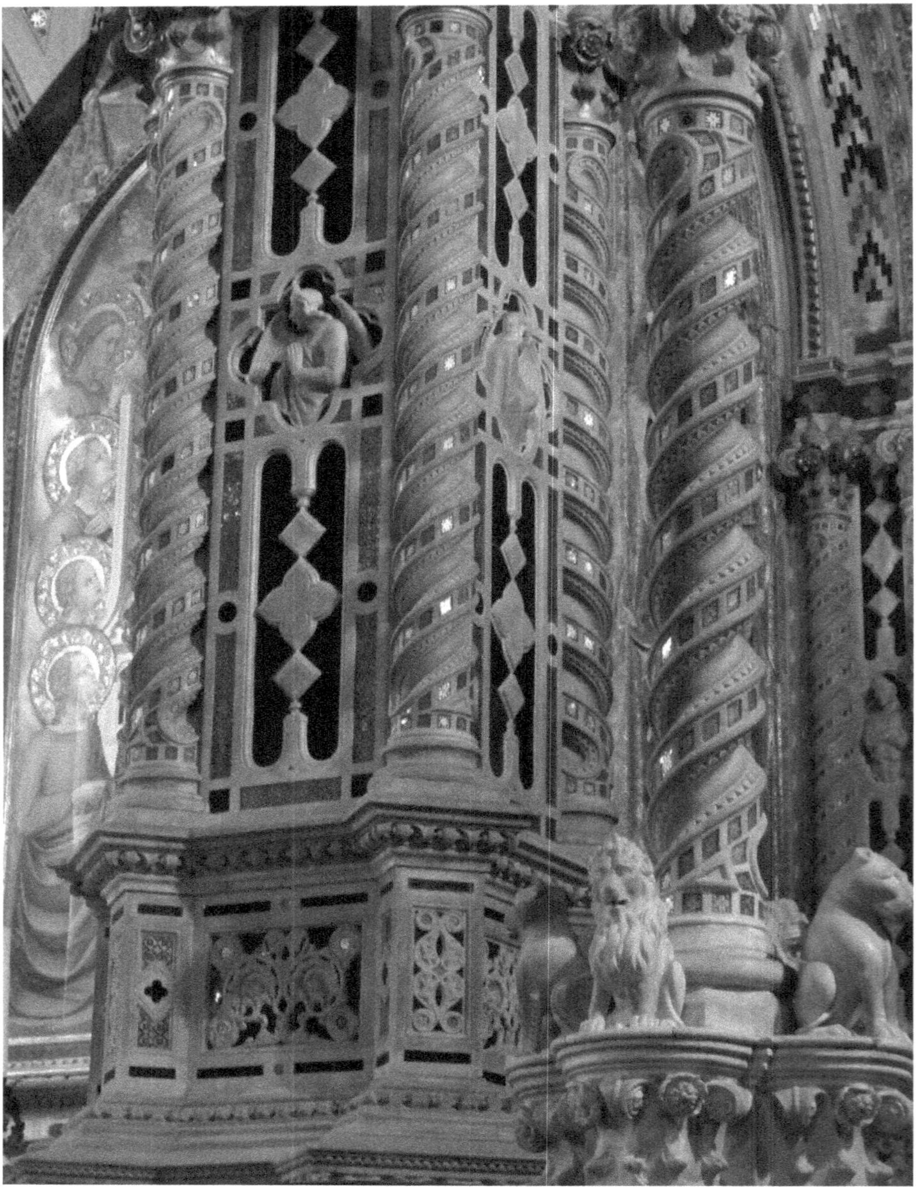

Figure 3.2: Orcagna, Detail of *Tabernacle* showing gilded glass, 1359, Orsanmichele, Florence. Source: Francesco Bini via Wikimedia Commons (CC BY-SA 3.0).

complex figurative imagery and added a higher degree of etched details than Greek artists. Romans frequently used gold-glass roundels featuring pagan, Jewish, or Christian themes to decorate the bottoms of bowls, cups, and other vessels (Figure 3.3).[31] Upon their owner's death, the gold-glass medallions were detached from the functional objects they decorated and impressed into the cement wall of the deceased's tomb in the catacombs. These examples are particularly important because they would have been accessible to medieval viewers when visiting the catacombs.

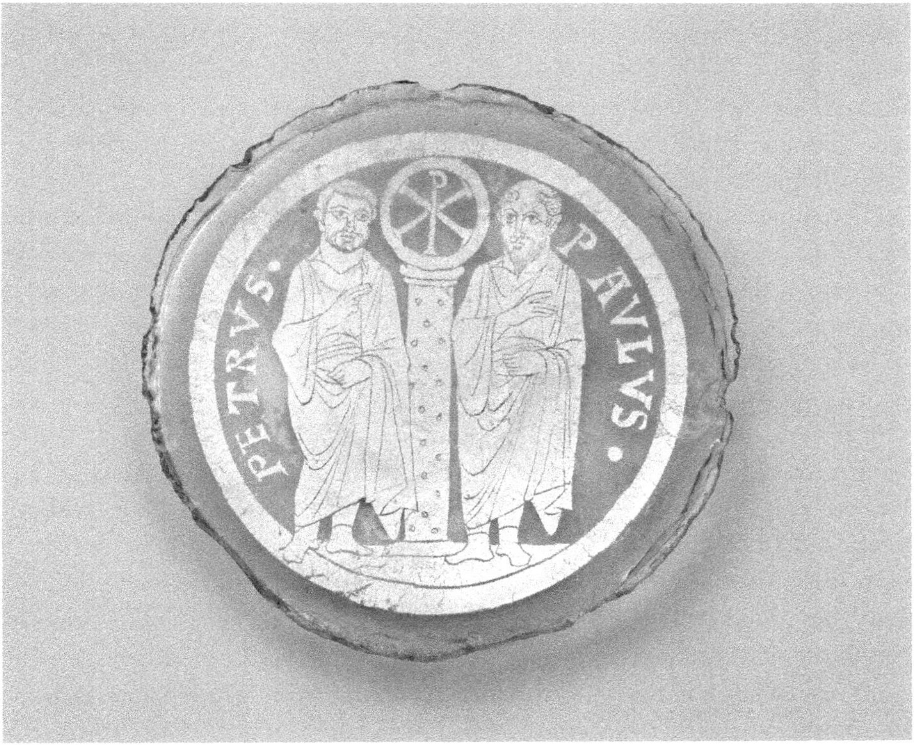

Figure 3.3: Byzantine Workshop, *Bowl Base with Saints Peter and Paul Flanking a Column with the Christogram of Christ*, late 4th century, The Metropolitan Museum of Art, New York, Rogers Fund, 1916 (16.174.3). Source: Public Domain, The Metropolitan Museum of Art (CC0).

Production of Roman gold glass peaked during the third and fourth centuries and then entered a period of sharp decline after the fifth century,[32] as did glass production in the Latin West more generally.[33] However, artists in Islamic territories, especially near Syria and around Egypt, continued to produce sandwich gold glass as well as many other types of glass, as will be discussed in more detail in Chapter 4. Characteristic examples of Syrian sandwich gold glass include the fragmentary cup in the David Collection of Copenhagen (inventory no. 4/1987), a bottle in the British Museum (inventory no. 1978,1011.2), and a cup in the Corning Museum of Glass (inventory no. 64.1.32). The production of gold-glass objects in Islamic territories was brief and limited compared with Roman output, being primarily confined to the ninth and tenth centuries.[34] Based on the relatively few extant examples and the short-lived nature of its production, Stefano Carboni suggests that Islamic gold glass was not commissioned by royal circles in a programmatic way but rather was an experimental response to the challenge of earlier Roman models.[35]

A resurgence of glass-making activities on the European continent occurred in the thirteenth century with the growing popularity of stained glass windows discussed in the last chapter. Gilded glass, too, experienced a revival, particularly in Italy with the technique of verre églomisé.[36] The Latin West's late-medieval revival of verre églomisé, a modified version of sandwich glass wherein the gold is backed with pigment rather than another sheet of glass, occurred in the later thirteenth century and was likely influenced by Roman examples.[37] Though Greeks and Muslims both made significant contributions to the history of gilded glass, duecento and trecento artists, patrons, and viewers would have had limited access to examples of Greek and Islamic sandwich gold glass. In contrast, Roman examples would have been easily accessible in terms of number and location. Furthermore, the Roman gold-glass roundels (Figure 3.3), or at least those associated with early Christians, could have held profound religious associations for later medieval viewers.

Some scholars have argued that these roundels served as a type of headstone or identification label, while others have argued that they functioned as protective devices, guarding the tomb of the deceased.[38] Although early Christians stopped using the catacombs as burial grounds after the fifth century, many Christians continued to visit the dead throughout the middle ages. Irina Taïssa Oryshkevich dispelled the previously held assumption that the catacombs went out of use during the later medieval and early Renaissance periods.[39] She supported her claim with literary evidence such as the *Mirabilia urbis Romae* and hagiographic texts in addition to archeological evidence such as the construction of aboveground cemeteries and churches that were physically linked to the catacombs by tunnel.[40]

She found that, despite the fact that many of the relics were removed to churches inside the city walls in the eighth century, the catacombs continued to be a source of spiritual and artistic inspiration for people of the later Middle Ages. And in some cases, the subterranean rooms were even turned into shrines.[41]

Visitors to the catacombs increased greatly around the year 1300 when Pope Boniface celebrated the Jubilee by issuing a papal bull that awarded pilgrims to Rome a spiritual reward on par with that offered to crusaders. In addition to an association with the holy sites of Rome, the ancient gold glass medallions may have also carried the aura of a secondary relic, that is, an object sanctified by physical contact with the remains of a holy figure. Not only did the gold-glass roundels press against the wall containing the remains of the deceased Christian martyrs, but while he or she was still living, the glass would have come into physical contact with the revered early Christian who had used it as a dish or vessel.

Evidence that a link between gilded glass and relics survived into early modern times is provided by Cennino Cennini's description of verre églomisé for his early modern treatise on artistic techniques, *The Craftsman's Handbook.* Before outlining the instructions for the technique, he writes that verre églomisé is a "process for working on glass, indescribably attractive, fine, and unusual, and this is a branch of great piety, for the embellishment of holy reliquaries."[42] The many extant reliquaries featuring verre églomisé indicate that fourteenth-century Italian artists in fact followed Cennini's advice. As most of these reliquaries also incorporate another type of glass, namely transparent glass, they will be discussed separately in Chapter 6.

Gilded glass, with its long association with divine illumination, its striking visual qualities, and its important artistic precedents, seems to have been ideally suited for demarking a space where the heavenly and earthly worlds converged. On the one hand, it is a solid, man-made, mundane material, and on the other, it produced a lighting effect that was elusive, immaterial, and supernatural. To investigate this premise further, the discussion that follows examines how specific artists, namely Nicola Pisano, Simone Martini, and Paolo di Giovanni Fei, each used the medium in a unique way, one that was suited to their specific context.[43]

Nicola Pisano's *Arca of San Domenico* and *Sienese Pulpit*

Nicola Pisano's pulpit for the Cathedral of Siena (1265–1268) and his *Arca of San Domenico* (1264–1267) are two of the earliest Italian monuments to incorporate gilded glass panels. Nicola and his workshop made the latter work, a monumental marble tomb to commemorate the revered founder of the Dominican order for the church dedicated to the saint in Bologna, in 1264.[44] Though Nicola's design was modified significantly from the fifteenth through eighteenth centuries, aspects of the original tomb are still visible. Modeled on ancient Roman art, as was typical of this artist's oeuvre, the sarcophagus features narrative relief sculptures showing episodes from the life of Saint Dominic on all four sides.

The figural groups fill most of the frame, leaving only small portions of a visible background (Figure 3.4). It is these small areas that are of greatest interest to this discussion because they are decorated with a series of small gilded glass panels. In a manner similar to the technique of verre églomisé, portions of the back of the panel are gilded—in this case the gilding is shaped into rosettes and other geometric shapes—and then the entire panel is backed with bright-red pigment. The result, when viewed from the front, reveals golden shapes set against a red background. When the lighting conditions were right, the gilded glass would have produced bright sparkling highlights visible from a significant distance. A closer viewer would have been struck by the prominent use of deep, almost blood-red pigment, a fitting color choice indeed for a burial monument.

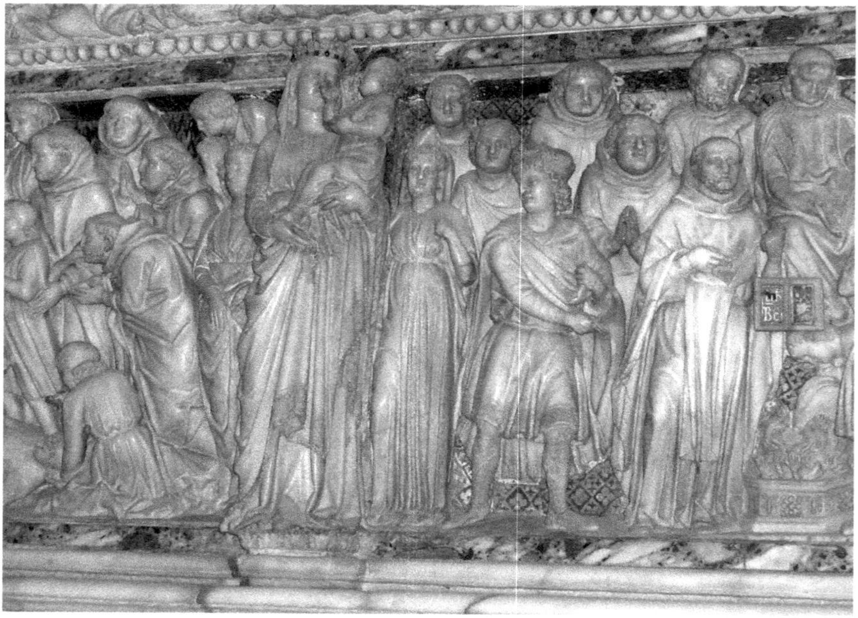

Figure 3.4: Nicola Pisano, *Arca of San Domenico,* 1264–1267, Basilica of Saint Dominic, Bologna. Source: Georges Jansoone via Wikimedia Commons (CC BY 2.5).

Nicola Pisano was both participating in a larger artistic dialog and forging new ground when it came to the way in which he used gilded glass. Certainly his approach should be situated alongside that found in mosaics, Cosmati, and early Christian roundels, as Nicola would have been familiar with at least some of these traditions. Similar to mosaics or Cosmati, the reflections created by the gilded glass in the *Arca* appear to strike out from the tomb toward the viewer's eyes. Whether or not he understood the visual theories of Plato or the Neoplatonists, his art—and his visual culture as a whole—was at least somewhat infused by the Byzantine tradition and its corresponding spiritual symbolism. The highlights of the *Arca* could have, therefore, resonated with a wider, more general appreciation for the "fiery rays" of the extramission visual theory and, in doing so, functioned as observable proof of the visual contact made between viewer and sacred monument.

Nicola's decision to use gilded glass in the *Arca* may have been influenced by older, established, connections between gold glass and relics. As will be discussed in greater detail in the chapters that follow, relics occupied a unique position within the Christian church. They were particularly sacred because of their uncanny ability to conjure aspects of a saint's earthly existence and their spiritual transcendence simultaneously. Relics paradoxically embodied an intersection of the spiritual and earthly worlds. In addition to the examples already mentioned—such as Cennini's recommendation, Orcagna's *Tabernacle*, and the tombs adorned with Cosmati—there were earlier, biblical precedents for using gold and glass to adorn holy places. According to Exodus 25, when God gave Moses directions to build the Ark of the Covenant, he ordered that gold be used extensively in its construction. The instructions are described as follows:

> You shall overlay [the ark] with pure gold, inside and out you shall overlay it, and you shall make a gold molding around it. You shall cast four gold rings for it and fasten them on its four feet, and two rings shall be on one side of it and two rings on the other side of it. You shall make poles of acacia wood and overlay them with gold. … You shall make a mercy seat of pure gold, two and a half cubits long and one and a half cubits wide. You shall make two cherubim of gold, make them of hammered work at the two ends of the mercy seat.

Interestingly, the directives for the ark do not mention gemstones, pigments, or other precious materials; the only decorative specifications describe the use of gold.

Other biblical references reinforce the notion that gold was an especially appropriate material for decorating a place of convergence between heaven and earth. King Solomon extensively used gold in the construction of his temple in a manner similar to the Ark of the Covenant. As it was described in 1 Kings 6:21–22,

> So Solomon overlaid the inside of the house with pure gold. And he drew chains of gold across the front of the inner sanctuary, and he overlaid it with gold. He overlaid the whole house with gold, until all the house was finished. Also the whole altar which was by the inner sanctuary he overlaid with gold.

Ultimately, both the temple and the ark were modeled on the most revered structure of all: Heavenly Jerusalem, which "shone with the glory of God, and its brilliance was like that of a very precious jewel, like a jasper, clear as crystal" and its street "was of gold, as pure as transparent glass," according to Revelations 21:9–21.

Gold—and by extension gilded glass—also had unusual physical qualities that may have contributed to its ability to adorn such sacred monuments. The appearance of gold was mutable; the very same piece of gold could look quite different depending on the environment. When lit by a strong light, a golden surface could shine with a bright yellow color, and without a strong light source, the same piece could appear to have darker, brownish hues and a more matte surface texture. Isidore of Seville comments on the fluctuating nature of the material in his *Etymologies* (ca. 615–630), noting how it "is named from 'gleam' (aura), that is from its luster, because it gleams more when the air reflects it" and "it is natural for the luster of metal to gleam more when it is reflected with another light."[45]

Gold's strong connection with divine symbolism may, therefore, stem from its changeable appearance and elusive nature, formal qualities that suggest aspects of immateriality. Lois Heidmann Shelton suggests that, for Byzantines, gold held special meaning because "within the hierarchy of the material and spiritual worlds, gold was the material closest to the immaterial and was connected by its luminosity, the source of its immaterial appearance, with the beauty and light of the Divine."[46] The gilded backgrounds of Byzantine icons were, therefore, crucial to their effectiveness as conduits of the divine because of their ability to connect the spiritual and material worlds, to give visible form to the invisible.

Western veneration of relics and the Eucharist was similar in nature to the Byzantine veneration of icons because these sacred treasures represented a tangible point of contact between heaven and earth. As the gold did for the icons, the gilded glass panels in the *Arca* brought visual form to the dichotomy between the heavenly and earthly with their sharp, glowing highlights. Thus, while gold's material value, beauty, formal qualities, and royal and divine associations may have all informed Nicola Pisano's decision to use it in the *Arca*, he also may have realized the ability of gilded glass to denote a convergence between the material and immaterial realms.

The interplay among the naturalistic sculptural details of Nicola's *Arca*, the reflective surface of the gilded glass, and the bright-red color of the pigment create a complex viewing experience with several layers of potential meaning. The highlights reflecting off the gold glass would have been a striking feature of the monument upon one's first impression. Even from a great distance, the bright, glittering light would have commanded the viewer's attention, inviting closer inspection of the sacred monument. As one approached the *Arca* and the sculptural details became clearer, the sparkling reflections may have competed with the figurative reliefs, at times even obscuring them, in a sense dematerializing them. As one strove to see and investigate the reliefs representing scenes from the saint's earthly existence, the glittering reflections persisted, and at the same time, the red pigment behind became more visible.

The bold use of red pigment for the *Arca* is conspicuous because no other extant work of Italian verre églomisé features it so prominently, nor does another artwork feature it at the exclusion of all other colors. When other verre églomisé works do use red, it is primarily reserved for the blood gushing from the crucified Christ's side. Thus, the contrast between the gilded glass and the sculptures creates a visual tension that mirrors the tension between Saint Dominic's human life and his saintly existence found in the relics. The relief panels tell the stories of Dominic's exemplary life, while the red pigment evokes the notion of the saint's blood, and the light from the gilded glass symbolizes the saint's heavenly afterlife.

Though art-historical scholarship has not yet addressed Nicola's inspiration or motivation for this innovative use of gilded glass, given his strong interest in classical art and the many ancient Roman gold-glass roundels found in the catacombs, it seems logical to posit that this could have been a contributing factor. In his biography of Nicola from the sixteenth century, Giorgio Vasari records that Nicola went to Viterbo and Naples. Although Vasari doesn't mention a trip to Rome, the Eternal City is located between these two cities and it would be hard to imagine an artist so interested in ancient Roman art missing a chance to visit there.

As mentioned previously, one of the most alluring destinations in Rome would have been the catacombs with their relic-like roundels of gold glass. Medieval Christian visitors making their way through the dark catacombs by candlelight would have found the gold-glass roundels and their glowing reflections visually striking and spiritually powerful. In such a setting, the gold glass demarked the location of important relics and in turn, the point of contact between the divine and the human spheres. As the physical remains or possessions of human beings who had become saintly spirits, relics belonged to both sacred and secular worlds simultaneously. Even if Nicola had not visited the catacombs in Rome, his strong interest in ancient Roman art made it likely that he would have been interested in other examples of it, for instance in the Camposanto of Pisa, which may have also had examples of gold glass.

Whatever his familiarity with or reason for using verre églomisé, Nicola must have found the medium effective, because he also used it in his next important commission. Only a year after receiving his commission for the *Arca*, Nicola and his workshop began work on the pulpit for the cathedral of Siena. This project was commissioned by the Opera of the Sienese Cathedral as part of a citywide improvement program, a program that reflected this city's cultural ascendancy as well as its fervent dedication to the Virgin Mary. Within the decade preceding work on the pulpit, Siena had enlisted and received the protection of the Virgin Mary to defeat the Florentines at the Battle of Montaperti in 1260 and completed construction on their cathedral, which was dedicated to the Madonna.

Similar to the *Arca*, the pulpit for the Cathedral of Siena included small panels of verre églomisé decorated with floral motifs set within the trilobed cusping and running along the base of the marble panels (Figure 3.5). Unlike the red pigment used in the *Arca,* here Nicola only used black paint to seal the gold leaf, and the amount of glass in the pulpit is significantly more limited. The verre églomisé embellishments certainly added an impressive lighting effect to the pulpit, but because it did not function as a tomb or reliquary, the symbolic interpretation assigned to gold in the discussion of Nicola's *Arca* does not readily apply. Furthermore, the pulpit in Siena features much less verre églomisé, and what gold glass is present is relegated to the base of the pulpit. It does not fill the background of the narrative panels. Therefore, upon initial comparison, the pulpit and *Arca* appear to have little in common. This is surprising, considering that the two works were made by the same workshop at about the same time.

When one considers the function of this structure, explanations for Nicola's differing treatment of the verre églomisé on his pulpit become more clear. Throughout the thirteenth and fourteenth centuries, priests would ascend such a pulpit in order to better address the congregation and read the Gospel or Epistles. However, this was not the only use for such a pulpit. The podium's height and visibility also made it the ideal place from which to display the church's prized relic collection.[47] Anita Moskowitz, in her discussion of the pulpit's use for the display of relics, notes that the relics of Saint Ranierus were displayed from a pulpit in 1161 in the cathedral of Pisa. Thus, like the tomb of Saint Dominic,

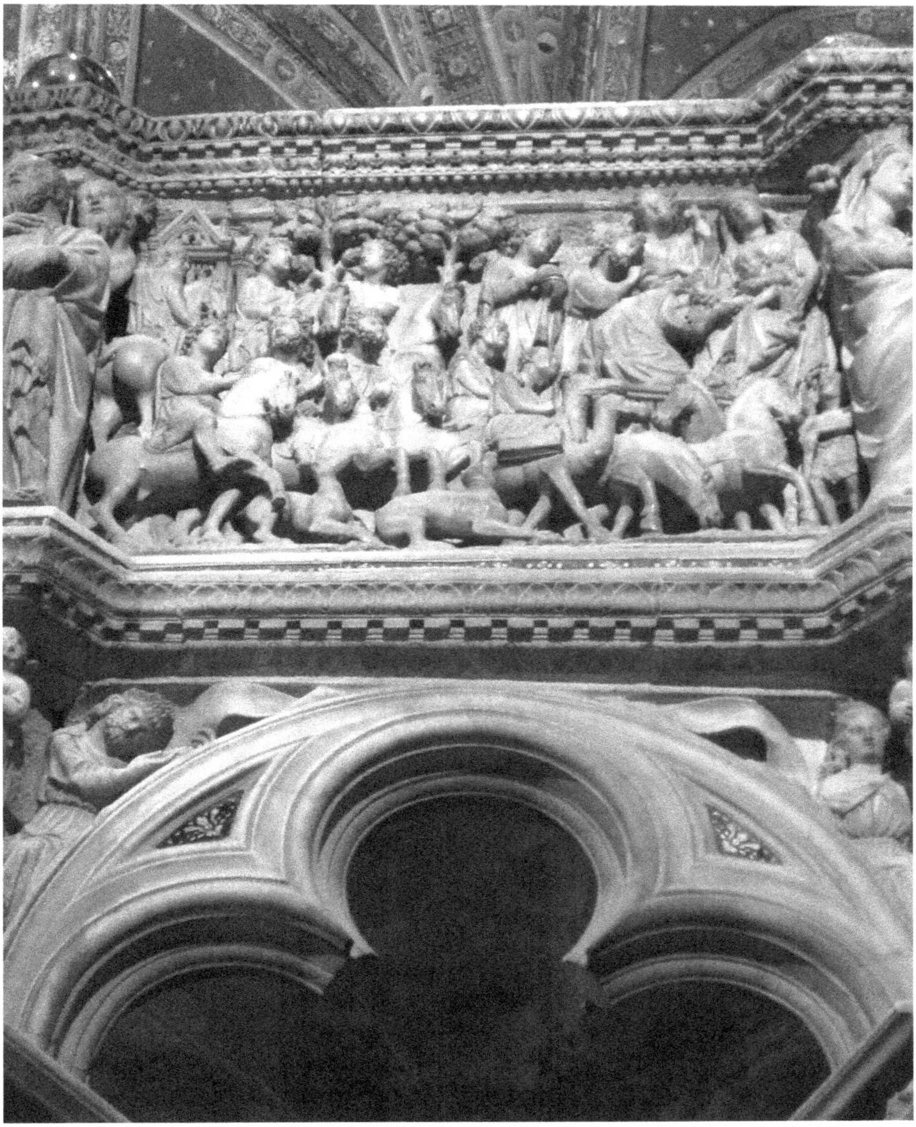

Figure 3.5: Nicola Pisano, Detail of *Pulpit* showing two glass panels on either side of the trilobed arch, 1265–1268, Siena Cathedral, Siena. Source: Author.

Nicola's pulpit could have also been associated with relics and therefore warranted a similar use of the verre églomisé. However, as the pulpit's primary function was preaching and the reading of the gospel, the use of the medium was significantly reduced.[48]

Therefore, Nicola Pisano may have used gilded glass to decorate areas associated with relics or evoke ideas related to divine illumination. In doing so, he followed the example of early Christian Romans, an artistic source he highly valued. Whether or not he was specifically aware of the extramission theory of vision and its fiery optical rays, he could still capitalize on the striking lighting effects created by the reflective gilded glass in order to create a powerful connection between viewer and his monument. Nicola Pisano's *Arca* and his pulpit for Siena's cathedral are not only instructive for their innovative adoption of gilded glass, but as two of the earliest major Italian monuments to incorporate the revived medium, they are also important case studies for an understanding of later usages of gilded glass, such as those by Simone Martini.

Simone Martini's *Maestà* and *Saint Louis of Toulouse*

As mentioned in the introduction chapter, Simone Martini used gilded-glass panels in his fresco of the *Maestà* in Siena's Palazzo Pubblico around 1315 and again in a panel painting commemorating Saint Louis of Toulouse around 1317. The *Maestà*, showing the enthroned Virgin Mary and Christ Child surrounded by a retinue of saints and angels, was painted on the wall of the Sala del Consiglio, a meeting chamber on the second floor of the city's town hall. Though much has been written about this work, most scholarly assessments of the fresco do not mention the approximately twenty-five panels of verre églomisé impressed into the wall's plaster, adorning the Madonna's throne and within Christ's halo (Figure 1.2).[49]

It is possible that Simone was inspired by Nicola Pisano's pulpit, which was located nearby in the town cathedral, only a short walk from the town hall. Not only was the cathedral a prominent artistic stage but Simone had worked on Duccio's great altarpiece for its high altar. This would have put him in direct contact with Nicola's gilded glass in the pulpit and also with the painted depiction of Cosmati in the Virgin's throne of Duccio's altarpiece. Simone's intimate familiarity with and high esteem for Duccio's altarpiece is evident in his fresco for the Palazzo Pubblico. Formal similarities suggest that Simone was, in fact, intentionally linking the two paintings in order to forge a connection between the two spaces.

Duccio *Maestà* for the cathedral and Simone's for the Palazzo Pubblico share strikingly similar formal qualities, especially in terms of composition. Both paintings feature rows of standing saints—each figure with individualized physiognomy and clothing—facing the Madonna and Child, while the patron saints of Siena kneel below and reverently gaze up at the Holy Family. Unlike Duccio's painting for the cathedral, however, Simone's *Maestà* for the civic center emphasizes the Virgin's secular power by surrounding her with earthly finery and depicting her as a queen. She wears a richly decorated blue garment bound by a brooch, which is not painted but made of an actual piece of glass, and sits in a sumptuous Gothic-style throne covered in gold with an elaborate cloth of honor hanging overhead. Simone's courtly painting style effortlessly combines sanctity and royalty. Thus Simone's painting diverges from Duccio's due to the different contexts in which these works were displayed; Simone's in a setting related to political power and Duccio's in a more religious setting.

Simone's use of verre églomisé can be seen as yet another way to solidify the connection between the city's religious center and political headquarters and yet simultaneously denote the different contexts of these important spaces. The glittering lighting effects seen in one space could have certainly reminded a viewer of the other, as could the appearance of the individual panels. Like Nicola's panels in the pulpit, Simone's verre églomisé panels are small, etched with gilded foliage motifs, and have black backing. They are also used relatively sparingly, at least compared with Nicola's *Arca*.

However, the gilded glass in the *Maestà* does not appear to signal the presence of relics; rather, it too seems to evoke an aura of courtly prestige. It is logical that Simone may have had different motivations when it came to gilded glass. He was working within a very different cultural moment from that of Nicola Pisano. Simone, living approximately a half of a century after Nicola, was therefore further removed from the Byzantine tradition and was operating in a context wherein civic and secular concerns were of paramount concern to his patrons; his first work with gilded glass was made for Siena's town hall and his second, a painting of Saint Louis of Toulouse, was made for the saint's brother, King Robert the Wise of Sicily.

Only two years after completing his *Maestà,* Simone used gilded glass with similar motivations and symbolism in mind in his panel commemorating Saint Louis of Toulouse. Simone used a piece of glass for the saint's morse, which features golden lilies and crosses to create the arms of Jerusalem and Sicily.[50] Simone's glass highlighted the royal heritage of Louis and his brother's divinely sanctioned authority by using verre églomisé for this important accessory. Thus, in both the *Maestà* and the panel of Saint Louis, verre églomisé adorned subjects associated with secular power, namely, the Madonna's throne and Saint Louis's coat of arms, both of which are related to notions of earthly power and a divinely sanctioned right to rule. Clearly the earthly status of these figures is not the only concern, however. Both the Madonna and Saint Louis are representatives of a heavenly court, not just an earthly one. Accordingly, Simone used gilded glass to symbolize the convergence of secular and sacred authority, following the example of Solomon's temple, a royal structure with strong divine connotations.

In contrast, when Simone's subject matter called for a more pious, spiritual emphasis, he did not use gilded glass. Simone painted a chapel dedicated to Saint Martin in the Lower Church of San Francesco in Assisi concurrently with his panel of Saint Louis. Here, Simone did not use any inset glass panels. The lack of gilded glass in the Chapel of Saint Martin becomes all the more conspicuous considering that every time Simone entered this chapel he would have passed underneath a visually striking example of silver-gilt glass. The ceiling of the Lower Church was painted dark blue with gold stars to resemble the sky, and to enhance the illusion of a night sky, artists set convex pieces of glass backed with silver into the contour of the painted stars.[51] When Simone entered the crypt-like lower church and made his way through the space towards his chapel, the silver-backed glass panels would have sporadically caught the light and appeared similar to twinkling stars in the night sky (Figure 2.5).

Simone, therefore, seems to have found verre églomisé better suited to subject matter with references to power and authority, but when it came to more solemn, humble subjects, he refrained. Simone was following a different tradition than that which inspired Nicola, albeit one with an equally long and well-established history. There were many reasons for gold's high value and its association with royalty throughout the middle ages. Unlike many other materials, in natural circumstances gold does not oxidize or otherwise deteriorate, earning its reputation as the most noble of all metals. This quality also made it ideal for use as currency, and its material worth meant it was typically reserved for either royal collections or royal burials.

As mentioned in the last chapter, Abbot Suger was also primarily concerned with royal patrons and used gold to embellish his royal commission at Saint Denis. An inscription by Suger over the main doors of this church once read:

> All you who seek to honor these doors,
> Marvel not at the gold and expense but at the
> craftsmanship of the work.
> The noble work is bright, but, being nobly bright, the work
> Should brighten the minds, allowing them to travel through
> the lights

> To the true light, where Christ is the true door.
> The golden door defines how it is imminent in these things.
> The dull mind rises to the truth through material things,
> And is resurrected from its former submersion when the
> light is seen.[52]

As the abbot instructed, one should not focus on the gold for its expense—or the wealth of the institution that funded it. Rather, one should appreciate the noble brightness of the work and see this as a reflection of Christ, the "true light." For Suger, then, the glittering golden surface of the doors was an analogy for divine enlightenment, not an indication of earthly splendor. However, the fact that he needed to explicitly instruct his viewers *not* to value the material itself may in fact reflect the tendency of a typical viewer to prize the worth of gold over its symbolism, especially considering the royal patronage of his commission.

As Simone Martini was working within the context of secular power structures, he needed to reference different associations from those of Nicola Pisano, even though the pulpit in the cathedral of Siena may have been a source of initial inspiration. The next case, Orcagna's *Tabernacle,* illustrates yet another approach shaped by a different set of circumstances entirely, namely the aftermath of the Black Death.

Orcagna's *Tabernacle* for Orsanmichele

Orcagna's *Tabernacle* for Orsanmichele, dating to 1352–1360 (Figure 3.2), closely follows the traditional Cosmati technique wherein the entire glass tessera is covered in a uniform sheet of gold leaf; there is no imagery etched into the gold. It was made approximately fifty years after Simone's fresco and panel, in response to the devastating bout of bubonic plague in 1348. The immense cost of Orcanga's monument was funded by donations and bequests after the plague. With so many citizens bequeathing funds to the church upon their deaths—or making donations to express their gratitude for surviving—abundant funds were available for a monument to protect the miracle-working image of the Madonna by Bernando Daddi. Indeed, the plethora of gilded glass, other precious materials, and sculptural reliefs indicates that Orcagna spared no expense when it came to decorating this marble monument.

While the influence of Simone Martini, Nicola Pisano, or even Giotto may have informed Orcagna's approach to gilded glass, another significant link is likely found in medieval Roman altar ciboria, which the artist could have seen on a

Jubilee trip to Rome in 1350. As discussed above, the technique of Cosmati originated in Rome, and if Orcagna participated in traditional Jubilee activities, such as making pilgrimages to various relic's shrines, he would have seen many examples of impressive Cosmati. Not only that, but he may have also been struck by how effective the ciborium format—a protected tabernacle composed of four columns and a pointed roof—was in both protecting and enshrining relics.[53] Thus, although Orcagna's structure was not a reliquary in the sense of containing actual relics, the miracle-working image it housed was, like relics, a physical, earthly object through which operated spiritual forces that needed to be visible and yet protected.

Much has been written exploring the Black Death's impact on early modern art, particularly on trecento art of central Italy, the primary topic of this study. Chapter 5 includes a more detailed discussion of the pertinent aspects of the debate as they apply to this book's thesis but a few words are instructive here as well. Millard Meiss's 1951 book, *Painting in Florence and Siena after the Black Death*, argued that the devastation caused by the plague was understood as divine retribution for society's growing interest in humanism and worldly interests and that this consequently resulted in a return to more conservative artistic trends and an abrupt break with naturalistic developments.

The trajectory of gilded glass presented in the present chapter appears to support Meiss's premise, at least to this point in the present discussion. Nicola Pisano's innovative gilded glass was used in conjunction with naturalistic sculptural reliefs inspired by ancient Roman art, while Simone Martini's glass adorns paintings that also display three-dimensionality, an interest in spatial perspective, and modeling of human figures. Thus, in both cases the artists were using verre églomisé in artworks that reflected a growing interest in the naturalistic style. In the first post-plague case study, namely Orcagna's *Tabernacle*, the artist has seemingly abandoned the verre églomisé technique in favor of the older style of Cosmati and appears to revert to a more straightforward adoption of the earlier examples wherein his glass is uniformly coated in gilding.

However, the rest of Orcagna's monument appears to deviate from the greater artistic conservatism that Meiss argues was inherent to the period. The relief sculptures, especially the narrative panels found along the base of the *Tabernacle*, display realistic three-dimensional space and figures. Furthermore, the centerpiece of the structure, that is the painting by Bernardo Daddi, also shows indications of naturalism in the arrangement of the angels on either side of the Madonna and the foreshortening of the throne. Following this thread, the next two case studies also seem to compromise Meiss's theory in that they are post-plague examples of artworks that feature verre églomisé along with highly innovative and naturalistic imagery.[54]

Paolo di Giovanni Fei and Lorenzo Monaco

Examples of gilded glass from later in the fourteenth century feature a striking departure from earlier models. Like the verre églomisé reliquaries which will be discussed in Chapter 6, artworks by Giotto's workshop (Figure 3.6), Paolo di Giovanni Fei (Figure 3.7) and Lorenzo Monaco treat the gilded glass almost like a canvas rather than a reflective element. In his panel of the *Madonna and Child* now in the Metropolitan Museum of Art, Paolo di Giovanni Fei inserted eleven verre églomisé medallions into the frame, two of which are now lost, and a larger one into the center of the panel to serve as the Madonna's brooch. In a manner very distinct from any of the previously discussed works, each of these small glass roundels features a modeled human face with detailed physiognomy and hairstyle, demonstrating the artist's keen interest in humanism and naturalism. The large roundel in the center of the painting features Christ, who looks directly outward toward the viewer and thus conveys a stoic, somber mood, while the saints depicted in the frame exhibit a livelier tone. The figures found in the frame, who have been identified as Peter, Paul, Matthew, John the Evangelist, John the Baptist, and Catherine of Alexandria, along with the Virgin Annunciate and the Angel Gabriel, appear to turn inward and gaze intently toward the Madonna and Child at the center of the painted panel.

Because these images are etched into the gold leaf, they have a sketch-like aesthetic. The linear quality makes them seem immediate and animated, as if the figures were caught in a specific moment or mid-gesture. As Silvana Pettenati observes, the graphic quality also, in a sense, makes them some of the only surviving drawings of the trecento.[55] Cennini seems to corroborate such a claim, as he too equated such imagery with the process of drawing. In his instructions for verre églomisé, he described how one should,

> Take a needle, fastened in a little stick as if it were a little brush, and have it quite sharp pointed. And, with the name of God, begin to draw lightly with this needle whatever figure you wish to make. And have this first drawing show very little, for it can never be erased; and therefore work lightly until you get your drawing settled; then proceed to work as if you were sketching with a pen, for this work has to be done freehand.[56]

Thus, like traditional drawings, these etched panels give the viewer a sense of the artist's gestures, and therefore their intimate, creative process. And yet, unlike

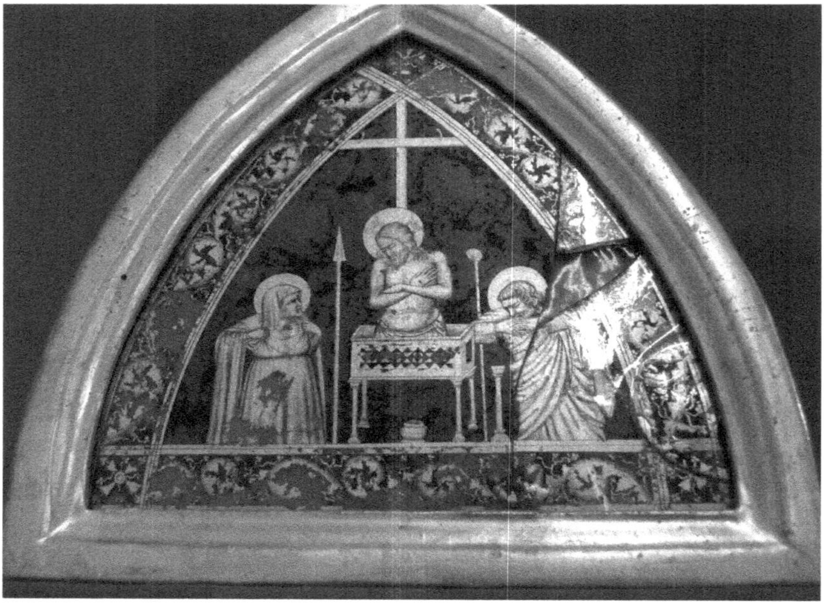

Figure 3.6: Workshop of Giotto, *Christ Surrounded by Mary and St. John*, ca. 14th century, The Bandini Collection, Fiesole. Source: Francesco Bini via Wikimedia Commons (CC BY-SA 3.0).

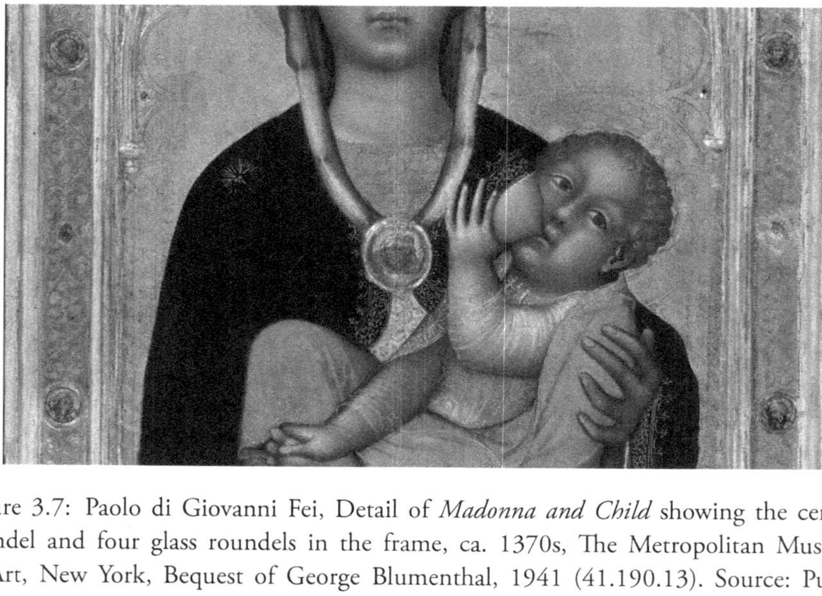

Figure 3.7: Paolo di Giovanni Fei, Detail of *Madonna and Child* showing the central roundel and four glass roundels in the frame, ca. 1370s, The Metropolitan Museum of Art, New York, Bequest of George Blumenthal, 1941 (41.190.13). Source: Public Domain, The Metropolitan Museum of Art (CC0).

traditional drawings, there is an aesthetic tension inherent in these images that does not allow the viewer to fully immerse themselves in the detailed naturalism.

The uneasy juxtaposition within the gold glass stems from the combination of two different types of lighting, referred to as surface light and pictorial light by Paul Hills in his influential book *The Light of Early Italian Painting.*[57] Hills defines surface light as the reflective highlights created by ambient light bouncing off smooth surfaces, and therefore this type of light is akin to the previously discussed concept of splendor. As previously discussed, the act of viewing of surface light was an experience that could have been interpreted as a powerful connection between the viewer and the holy image, as light and spiritual enlightenment were commonly associated with each other. In his discussion of gilded paintings, Hills notes how "the light of Christ's brightness could literally shine upon the eyes; and this physical light entering the eyes would have been intuitively understood as metaphor for spiritual illumination."[58] Though Hills discusses panel paintings, the same concept could certainly have applied to the many examples of gilded glass under discussion in this chapter, and furthermore, such an analysis supports the connection between Franciscan light theology and the artistic uses of gilded glass discussed previously.

Pictorial light, on the other hand, is defined by Hills as the illusionistic lighting within a work of art that expressed mass and depth through the description of highlights and shadows. In the glass roundels by Paolo di Giovanni Fei, instances of pictorial light are found within the highlights and shadows of the figures. More specifically, pictorial light is evident in the modeling that defines the Virgin Annunciate's cheeks, chin, and neck. As she recoils from Gabriel's message, the side of the Virgin's face farthest from the viewer is heavily shadowed. In a similar way, the areas just below her eyes, lips, and chin are also modeled with shadow to create depth.

Lorenzo Monaco engaged with similar techniques in his *Madonna of Humility with Two Saints* from 1408 in the Museo Civico of Turin and the extant fragment of *The Virgin and Child Enthroned with Saint John the Baptist and Saint John the Evangelist* from the late fourteenth century in the Louvre's collection.[59] In both these verre églomisé panels, the imagery inscribed into the gold leaf is highly modeled, detailed, and spatially complex. Adding an even more painterly quality to the works is the fact that they use different colors throughout the backing pigment. In the Louvre's fragmented panel, for instance, the Madonna wears a red dress and blue mantle as she sits before a red cloth of honor that features a complicated organic pattern. The flanking figures of John the Baptist on the left and what is likely John the Evangelist on the right, are equally impressive; in fact, their red draped clothing, facial expressions, and poses are closer to finished paintings than the sketch-like aesthetic of the roundels by Paolo di Giovanni Fei.

The interdependence of surface light and pictorial light within the gilded glass of Paolo di Giovanni Fei and Lorenzo Monaco is unusual in late medieval art as a whole. Not many artistic techniques simultaneously blend such a high degree of naturalism with such reflective surface sheen. One technique worth considering here is the depiction of figures, and even narrative scenes, found in enamel works. For instance, the scenes depicted in the enamel roundels of the *Stavelot Triptych* feature overlapping figures, which creates a sense of depth, and the clothing worn by the figures suggests physical bodies beneath that bend, twist, and show through the drapery. But such enamel scenes lack the subtle modulation of highlights and shadows found in the glass panels by Lorenzo Monaco or Paolo di Giovanni Fei where the physicality of the modeled faces is inextricably bound to the supernatural sparkle of the gold leaf.

In most other artworks and techniques, surface light and pictorial light lie on a continuum, and as Paul Hills observes, there was a general shift away from interest in surface light, which he saw as characteristic of the medieval aesthetic, toward a greater interest in illusionistic lighting effects, which he understood as characteristic of the Renaissance style.[60] As this chapter has shown, however, much like Meiss's theory on post-plague art, this paradigm does not readily apply to these specific examples of gilded glass.

Conclusions on Gilded Glass: Harnessing Divine Light and Fiery Rays

Some of the abovementioned artworks—like Simone Martini's *Maestà*—are famous and well-studied commissions of late medieval and early Renaissance Italy, which when they are contextualized within the development of gilded glass, reveal new insights about these artists and this time period. Other works—like Paolo di Giovanni Fei's panel—are less famous works that present significant questions for our understanding of major trends commonly associated with this historical period. Inspired by earlier works of art with gold and glass, late medieval and early Renaissance gilded glass—both with and without etched imagery—had the ability to create visually impressive and symbolically rich effects, but the exact nature of this meaning took different forms depending on the work's context and the way in which the artist used verre églomisé.

For example, in the work of Simone Martini, the gilded glass brought earthly splendor and spiritual authority to the Virgin and Saint Louis by illuminating their symbols of civic-spiritual power, namely the Heavenly Virgin's throne and the arms of Sicily and Jerusalem. For Nicola Pisano on the other hand, the use of gilded glass correlated with the presence of holy relics. And yet in each case, the artist may have been capitalizing on the visual effects of gilded glass to signify a profound moment of visual contact and may have perhaps even understood the sparkling reflections as visual analogies for Plato's fiery rays.

By combining gold and glass with other artistic media, each of the artists discussed here also created works of art that resonated with a more abstract outlook and at the same expressed an interest in a new, more naturalistic approach to the visual arts. This feature is most evident in the cases of Paolo di Giovanni Fei and Lorenzo Monaco, but it is also found in the work of Nicola Pisano, Simone Martini, and Orcagna, each of whom incorporated gilded glass into their works with naturalism. Nicola Pisano's monuments display some of the most innovative *all'antica* sculptural reliefs of their time, while Simone Martini's paintings and Orcagna's reliefs all evoke realistic, three-dimensional forms and space.

Furthermore, the theme of duality is a reoccurring one when discussing artworks with gilded glass. This is most characteristic in Nicola Pisano's *Arca di San Domenico,* which created bright, golden reflections that dematerialized the monument's surface and at the same time evoked a more earthly, material nature through the plasticity of the relief sculptures. The way Nicola used gilded glass eloquently evoked the dual nature of the relics—which are both material and divine objects—and ultimately, the dual nature of Christ as both human and God. The physical properties of glass may have contributed to this unique ability. Gilded glass had several biblical references and rich religious symbolism, and yet it was known to be a man-made substance that began as lowly sand and ash. Much more will be said about such themes in Chapter 6.

The case studies here also contribute to one of this book's overarching themes, which was discussed at the end of the last chapter, namely that certain artists operated within a network of glass-related ideas and tended to use the medium in a variety of ways that reflect a conscious exploration of its optical associations. Already in the first two chapters it has become clear that Simone Martini was one such artist. In his chapel of Saint Martin, Simone explored innovative ways of forging meaningful connections between the chapel's stained glass windows and its frescoes, and as discussed in this chapter, the same artist incorporated panels of gilded glass into one of his most prestigious panel paintings and one of his most prominent frescoes. Giotto, too, was exploring various types of glass. As mentioned in the last chapter he, or his workshop, depicted what appear to be mirrors in the pinnacle of the Baroncelli Altarpiece, he used actual gilded glass in the Crucifixion panel for Santa Maria Novella, and he depicted gold glass Cosmati in the *Stefaneschi Altarpiece.* The next few chapters consider yet another type of glass, namely transparent glass, and in doing so, even more connections will be revealed.

Notes

1. For general information on these techniques, see Ferdinando Rossi, *Mosaics: A Survey of Their History and Techniques* (New York: Praeger, 1970); Janetta Rebold Benton, "Mosaic: Sparkling Surfaces," in *Materials, Methods, and Masterpieces* (Santa Barbara: Praeger, 2009); Edward Hutton, *The Cosmati: The Roman Marble Workers of the Twelfth and Thirteenth Centuries* (London: Routledge and Paul, 1950); and Paloma Pajares-Ayuela, *Cosmatesque Ornament: Flat Polychrome Geometric Patterns in Architecture* (New York: Norton, 2001).
2. An important exception includes Marco Ciatti and Max Seidel, eds, *Giotto: The Santa Maria Novella Crucifix*, trans. Mark Roberts and Svitlana Claudia Hluvko (Florence: Edifir, 2002).
3. Claus-Christian Carbon, "Golden perception: Simulating Perceptual Habits of the Past." *i-Perception* 4 (2013): 468–76; Claus-Christian Carbon, "The Cycle of Preference: Long-term Dynamics of Aesthetic Appreciation," *Acta Psychologica* 134, no. 2 (June 2010): 233–44; Ute Leonards et al. "Mediaeval Artists: Masters in Directing the Observers' Gaze," *Current Biology* 17, no. 1 (2007): R8–R9; Margaret Livingstone, *Vision and Art: The Biology of Seeing* (New York: Abrams, 2002); Claudia Muth and Claus-Christian Carbon, "The Aesthetic Aha: On the Pleasure of Having Insights into Gestalt," *Acta Psychologica*, 144, no. 1 (September 2013): 25–30; Wolfgang Schöne, *Über das Licht in der Malerei [About the Light in Painting]* (Berlin: Mann, 1954); Hans Sedlmayr, *Über Farbe, Licht und Dunkel [On Colour, Light and Darkness]* (Munich: Hueber, 1959); Robert S. Nelson, *Visuality Before and Beyond the Renaissance: Seeing as Others Saw* (Cambridge: Cambridge University Press, 2000); and Paul Hills, *The Light of Early Italian Painting,* (New Haven, CT: Yale University Press, 1987).
4. Carbon, "Golden Perception," 475.
5. Ibid.
6. Hills, *Early Italian Painting,* 11. For more discussion on theories of the nature of light at this time, also see Chapter 6.
7. Bonaventure, "The Soul's Journey into God," in *Bonaventure: The Soul's Journey into God, the Tree of Life, the Life of Saint Francis*, ed. Ewert Cousins (New York: Paulist Press, 1978), 53.

8. Bonaventure, "The Tree of Life," in Cousins, *Bonaventure*, 170–71. For an interpretation of the emanation of light as the divine Logos see Hills, *Early Italian Painting*, 14.
9. Arthur Zajonc, *Catching the Light: The Entwined History of Light and Mind* (Oxford: Oxford University Press, 1993), 97–98; Edward Grant, ed., *A Source Book in Medieval Science* (Cambridge, MA: Harvard University Press, 1974), 383.
10. Hills, *Early Italian Painting*, 11.
11. Hills, *Early Italian Painting*, 11 and Nancy Thompson, "The Franciscans and Stained Glass in Tuscany and Umbria", in *Mendicant Cultures in the Medieval and Early Modern World: Word, Deed, and Image*, eds. Sally J. Cornelison, Nirit Ben-Aryeh Debby, and Peter Howard (Turnhout: Brepols, 2015), esp. 27–32.
12. Hills, *Early Italian Painting*, 11.
13. For more detail on these insets, see Ciatti and Seidel, *Giotto*, 53.
14. His inspiration may have been the Master of the Franciscan Crucifix. See Silvana Pettenati "The Decorated Glass," in Ciatti and Seidel, *Giotto*, 209.
15. Hills, *Early Italian Painting*, 11. For more on Bartolomeo da Bologna (sometimes written Bartolomeo di Bartoli da Bologna, Bartholomew of Bologna, or Bartholomaei de Bononia), see Graziella Federici-Vescovini, *Le teorie della luce e della visione ottica dal IX al XV secolo: Studi sulla prospettiva medievale e altri saggi* (Perugia: Morlacchi Editore, 2003), 29–33.
16. Irenaeus Squadrani, "Tractatus de luce Fr. Bartholomaei de Bononia," *Antonianum* 7 (1932): 230–31.
17. A similar interpretation of light reflecting off gold as seen in the medieval tradition can be found in what Paul Hills describes as surface light. Hills, *Early Italian Painting*, 3–28.
18. David C. Lindberg, *Theories of Vision from Al-kindi to Kepler* (Chicago: University of Chicago Press, 1976), 5.
19. Carolyn L. Connor, *Saints and Spectacle: Byzantine Mosaics in Their Cultural Setting* (New York: Oxford University Press, 2016), 135.
20. Isidore of Seville, The Etymologies VI.ix. 2–xii. 3 in Stephen A. Barney et al., *The Etymologies of Isidore of Seville* (New York: Cambridge University Press, 2006), 141.
21. For general texts on these techniques, see Ferdinando Rossi, *Mosaics: A Survey of Their History and Techniques* (New York: Praeger, 1970); Benton, "Mosaic;" Hutton, *Cosmati*; Pajares-Ayuela, *Cosmatesque Ornament*; Silvana Pettenati, *I vetri dorati graffiti e i vetri dipinti* (Turin: Museo Civico di Torino, 1978), xv–xlviii.
22. Erkinger Schwarzenberg, "Colour, Light, and Transparency in the Greek World," in *Medieval Mosaics: Light, Color, Materials*, eds. Eve Borsook, Fiorella Gioffredi Superbi, and Giovanni Pagliarulo (Florence: Silvana, 2000), 29. Photismos has been translated as *the act of enlightenment, illumination* or *light*.
23. Eve Borsook, "Rhetoric or Reality: Mosaics as Expressions of Metaphysical Idea," *Mitteilungen des Kunsthistorischen Institutes in Florenz* 44 (2000): 9.

24. Borsook, "Rhetoric or Reality," 3. For more on mosaic inscriptions, see Erik Thunø, "Inscription and Divine Presence: Golden Letters in the Early Medieval Apse Mosaic," *Word and Image* 27, no. 3 (2011): 279–91.
25. Eve Borsook, introduction to Borsook, Superbi, and Pagliarulo, *Medieval Mosaics*, 13–14.
26. See Connor, *Saints and Spectacle*, 132, which cites Leslie Brubaker, "Byzantine Art in the Ninth Century: Theory, Practice, and Culture," *Byzantine and Modern Greek Studies* 13, no. 1 (1989): 66–67.
27. Borsook, introduction to Borsook, Superbi, and Pagliarulo, *Medieval Mosaics*, 13.
28. Other notable sites in Rome featuring the medium: San Paolo fuori le Mura, Santa Maria in Aracoeli, Santa Maria in Cosmedin, Santa Maria Maggiore, and the Stanza della Segnatura.
29. Hugh Tait, ed., *Five Thousand Years of Glass* (London: British Museum Press, 1991), 49. For more on Hellenistic sandwich glass, see Jennifer Price, "Glass," in *A Handbook of Roman Art,* ed. Martin Henig (London and New York: Phaidon Press, 2006), 206. Though it doesn't use transparent glass, an Etruscan earring from the sixth century BCE displays a similar technique: gold filigree is covered by a piece of rock crystal. "Origin and Influence, Cultural Contacts: Egypt, the Ancient Near East, and the Classical World," *Metropolitan Museum of Art Bulletin* 29 no. 7 (March 1971): 326.
30. Price, "Glass," 207.
31. For more on the Early Christian use of verre églomisé, see C. Louise Avery, "Early Christian Gold Glass," *Metropolitan Museum of Art Bulletin* 16 no. 8 (August 1921): 170–75; O. M. Dalton, "The Gilded Glasses of the Catacombs," *Archaeological Journal* 58 (1901): 225–53.
32. Ibid., 171; Pettenati, *I vetri dorati*, 17.
33. Alan Macfarlane and Gerry Martin, *Glass: A World History* (Chicago: University of Chicago Press, 2002), 16–18.
34. Stefano Carboni and David Whitehouse, *Glass of the Sultans* (New York: Metropolitan Museum of Art, 2001), 222.
35. Ibid.
36. For more on the technique's history, see Pettenati, *I vetri dorati*, xv–xlviii.
37. Another variation featured painted glass backed with gold leaf. See Pettenati, "Decorated Glass," 209 in Ciatti and Seidel, *Giotto*, 209.
38. C. Louise Avery discussed the traditional theory that they were used for identification purposes as well as Eisen's theory that they served a protective capacity, warning evil spirits to stay away from the Christians. For more on this debate and for further bibliography, see C. Louis Avery, "Early Christian Gold Glass," *The Metropolitan Museum of Art Bulletin* 16 no. 8 (August 1921): 170.
39. Irina Taïssa Oryshkevich, "The History of the Roman Catacombs from the Age of Constantine to the Renaissance" (PhD diss., Columbia University, 2003).

40. For more on tunnels from churches to catacombs, see ibid., 17. See also Laurie Brink and Deborah Green, *Commemorating the Dead: Texts and Artifacts in Context; Studies of Roman, Jewish, and Christian Burials* (New York: Walter de Gruyter, 2008), 14.
41. Debra Julie Birch, *Pilgrimage to Rome in the Middle Ages: Continuity and Change* (Woodbridge, UK: Boydell Press, 1998), 13.
42. Cennino Cennini, *Il libro dell'arte*, trans. Daniel V. Thompson Jr. (New Haven: Yale University Press, 1933). See text on page 107 and the translation on page 112. For Italian text, see also Carlo Milanesi and Gaetano Milanesi, eds., *Il libro dell'arte o trattato della pittura, di Cennino Cennini da Colle Valdelsa; di nuovo pubblicato con molte correzioni e coll'aggiunta di più capitoli tratti dai codici fiorentini* (Florence: Felice Le Monnier, 1859), 123.
43. This list is not meant to be exhaustive, rather a sampling of prominent artists and artworks. More work is needed on the wider network of artists working with glass. For instance, the *Tomb of Mary of Hungary* by Tino di Camaino in Santa Maria Donna Regia in Naples from around 1325 is another interesting example of an artist using gilded glass in a tomb.
44. According to Anita Moskowitz, the earliest sources on the *Arca* do not name the sculptor, but it has long been associated with Nicola Pisano because of Vasari's attribution and based on stylistic reasons. Anita Fiderer Moskowitz, *Nicola Pisano's Arca di San Domenico and Its Legacy* (University Park, PA: Pennsylvania State University Press, 1994), 8. For more on the dating and documentation see ibid. For a more detailed description of Nicola's work see ibid., 9.
45. Barney, *Isidore of Seville*, 329.
46. Lois Hiedmann Shelton, "Gold in Altarpieces of the Early Italian Renaissance: A Theological and Art Historical Analysis of Its Meaning and of the Reasons for Its Disappearance" (PhD diss., Yale University, 1987), 25.
47. Moskowitz, *Arca di San Domenico*, 21. For more on the use of pulpits for the display of relics and the connections between Pisano's treatment of his pulpits and the *Arca*, see ibid.
48. The fact that one of Nicola's assistants used verre églomisé in a similar way on another pulpit only a few years later is further evidence that Nicola's work was not an anomaly. In approximately 1270, Guglielmo da Pisa—one of Nicola's assistants on the *Arca*—applied a similar treatment to that on the saint's tomb to his pulpit for San Giovanni Fuorcivitas in Pistoia, a fragment of which is preserved in that city's Diocesan Museum; see Stefano Carboni, *Venice and the Islamic World, 828–1797* (New York: Metropolitan Museum of Art, 2007), 255–56. Carboni suggests that the pattern in this piece of glass and perhaps even the use of verre églomisé was inspired by Islamic examples of sandwich glass from ninth- and tenth-century Syria, an issue which is discussed in more detail in Chapter 4. In the catalogue entry for the fragment of glass (Carboni, *Venice*, 340–41) Carboni follows Bertelli's attribution to

Venetian craftsmen working in Pistoia, citing the Venetians' close contact with Muslims. But Carboni's theory does not take into consideration that Nicola Pisano used it and Guglielmo could have seen it in his master's works.

The verre églomisé is found in the background of the reliefs and likely lined the bottom of the pulpit and the bookstand. It therefore would have covered a much larger area than that found in Nicola's pulpit. A possible explanation of Guglielmo's more extensive treatment may relate to a myth demonstrating his deep personal passion for relics, found in the Necrologio of the Convent of Santa Caterina, Pisa, written by Fra Domenico da Peccioli (d. 1407). The story tells how, when Guglielmo attended the translation of Saint Dominic's relics, he was so overcome with reverential passion that he stole a relic of the saint's rib for his convent of Santa Caterina; Moskowitz, *Arca di San Domenico*, 8.

49. John White, *Art and Architecture in Italy: 1250–1400* (New Haven, CT: Yale University Press, 1993), 349–50.
50. Julian Gardner, "Saint Louis of Toulouse, Robert of Anjou and Simone Martini," *Zeitschrift für Kunstgeschichte* 39 (1976): 12.
51. Hayden Maginnis, "Assisi Revisited: Notes on Recent Observations," *The Burlington Magazine* 117, no. 869 (August 1975): 512.
52. Cited in Erwin Panofsky, *Abbot Suger on the Abbey Church of St. Denis and Its Art Treasures* (Princeton, NJ: Princeton University Press, 1979), 47–49.
53. Like earlier scholars, Brendan Cassidy finds parallels between the work of Arnolfo di Cambio and Orcagna, but Cassidy differs in that he maps out a series of other, specifically reliquary structures as well. Brendan Cassidy, "*Orcagna's Tabernacle* in Florence: Design and Function," *Zeitschrift für Kunstgeschichte* 55 (1992): 180–211.
54. There are also many examples discussed in Chapter 6 that provide evidence that artists making verre églomisé continued to experiment with innovation and naturalism throughout the fourteenth century.
55. This keen observation was made by Silvana Pettenati in "The Decorated Glass," in Ciatti and Seidel, *Giotto*, 209. Although she was not specifically referring to the works by Paolo di Giovanni Fei, her insight can also apply to this case.
56. The full set of instructions is as follows:

> Take a piece of white glass, with no green cast, very clean, free from bubbles; and wash it, rubbing it down with lye and charcoal. And rinse it with good clear water, and let it dry by itself. But before you wash it, cut it to the size you want. Then take the white of a fresh egg; beat it with a good clean whisk just as you do that for gilding, so that it is thoroughly beaten; and let it distil overnight. Then take a minever brush, and with this brush wet the back of the glass with this glair; and when it is thoroughly wet all over, take a leaf of the gold, which should be quite heavy gold, that is, dull; put it on the paper tip, and lay it deftly on the glass where you have wet it; and press it down with a little very clean cotton, gently, so that the glair does not get on top of the gold; and lay the whole glass in this way. Let it dry without sun for the space of some days.

> When it is all dry, get a nice flat little panel, covered with black cloth or silk; and have a little study of your own, where no one will cause you any sort of interruption, and which has just one cloth-covered window; and you will put your table in this window, as if for writing, so arranged that the window shines over your head when you have your face turned toward this window. With your glass laid out on this black cloth:
>
> Take a needle, fastened in a little stick as if it were a little brush, and have it quite sharp pointed. And, with the name of God, begin to draw lightly with this needle whatever figure you wish to make. And have this first drawing show very little, for it can never be erased; and therefore work lightly until you get your drawing settled; then proceed to work as if you were sketching with a pen, for this work has to be done freehand. And do you want to be convinced that you need to have a light hand, and that it should not be tired?—[Know] that the strongest shadow you can make consists in penetrating to the glass with the point of the needle, and no more; that the intermediate shadow consists in not piercing through the gold all over; that it is as delicate as that, and you must not work with haste—rather with great enjoyment and pleasure. And I give you this advice, that the day before the day you want to work at this job, you hold your hand to your neck, or in your bosom, so as to get it all unburdened of blood and weariness.
>
> When you have got your drawing finished, and you want to scrape away certain grounds, which generally want to be put in with ultramarine blue in oil, take a leaden style, and rub the gold, which it takes off for you nearby; and work carefully around the outlines of the figure. When you have done this:
>
> Take various colors ground in oil, such as ultramarine blue, black, verdigris, and lac; and if you want any drapery or lining to glisten [in lines of gold] on green, apply green; if you want it on lac, apply lac; if you want it on black, apply black. But the black is the most striking of all, for it shows up the figures better than any other color.

Cennino Cennini, *Il libro dell'arte*, trans. Daniel V. Thompson Jr. (New Haven: Yale University Press, 1933). See text on pages 107–8 and the translation on pages 112–14. For Italian text, see also Carlo Milanesi and Gaetano Milanesi, eds., *Il libro dell'arte o trattato della pittura, di Cennino Cennini da Colle Valdelsa; di nuovo pubblicato con molte correzioni e coll'aggiunta di più capitoli tratti dai codici fiorentini* (Florence: Felice Le Monnier, 1859), 123–24.

57. Hills, *Early Italian Painting*, 3–28.
58. Ibid., 18. Hills also writes, "the lustre moving over the surface … could be experienced as a personal link between the worshipper's eyes and the sacred image."
59. For more on the Louvre's panel, see Laurence B. Kanter et al., *Painting and Illumination in Early Renaissance Florence, 1300–1450* (New York: Harry N. Abrams, 1994), 223–26; and for more on the panel in Turin, see Pettenati, *I vetri dorati*, 15–17.

60. Hills is clear that this was not the only time an interest in illusionistic space was demonstrated. For more on his observations of an interest in naturalistic space in ancient times, see Hills, *Early Italian Painting*, 9–10.

References

Avery, C. Louis. "Early Christian Gold Glass." *The Metropolitan Museum of Art Bulletin* 16, no. 8 (August 1921): 170–75.

Barney, Stephen A., W. J. Lewis, J. A. Beach, Oliver Berghof, and Muriel Hall, eds. *The Etymologies of Isidore of Seville*. Cambridge: Cambridge University Press, 2006.

Benton, Janetta Rebold. *Materials, Methods, and Masterpieces of Medieval Art*. Santa Barbara: Praeger, 2009.

Birch, Debra Julie. *Pilgrimage to Rome in the Middle Ages: Continuity and Change*. Woodbridge, UK: The Boydell Press, 1998.

Bonaventure. "The Soul's Journey into God." In *Bonaventure: The Soul's Journey into God, The Tree of Life, The Life of St. Francis*, ed. Ewert Cousins, 51–117. New York: Paulist Press, 1978.

———. "The Tree of Life." In *Bonaventure: The Soul's Journey into God, The Tree of Life, The Life of St. Francis*, ed. Ewert Cousins, 118–168. New York: Paulist Press, 1978.

Borsook, Eve. *Medieval Mosaics: Light, Color, Materials*. Milan: Silvana, 2000.

———. "Rhetoric or Reality: Mosaics as Expressions of Metaphysical Idea." *Mitteilungen des Kunsthistorischen Institutes in Florenz* 44 (2000): 3–18.

Brink, Laurie, and Deborah Green. *Commemorating the Dead: Texts and Artifacts in Context: Studies of Roman, Jewish, and Christian Burials*. New York: Walter de Gruyter, 2008.

Brubaker, Leslie. "Byzantine Art in the Ninth Century: Theory, Practice, and Culture." *Byzantine and Modern Greek Studies* 13, no. 1 (1989): 66–67.

Carbon, Claus-Christian. "The Cycle of Preference: Long-term Dynamics of Aesthetic Appreciation," *Acta Psychologica* 134, no. 2 (June 2010): 233–44.

———. "Golden perception: Simulating Perceptual Habits of the Past." *i-Perception* 4 (2013): 468–76.

Carboni, Stefano, and David Whitehouse. *Glass of the Sultans*. New York: Metropolitan Museum of Art, 2001.

———. *Venice and the Islamic World: 828–1797*. New York: Metropolitan Museum of Art, 2007.

Cassidy, Brendan. "*Orcagna's Tabernacle* in Florence: Design and Function." *Zeitschrift für Kunstgeschichte* 55 (1992): 180–211.

Cennini, Cennino. *Il libro dell'arte*. Translated by Daniel V. Thompson, Jr. New Haven, CT: Yale University Press, 1933.

Ciatti, Marco and Max Seidel, eds. *Giotto: The Santa Maria Novella Crucifix*. Translated by Mark Roberts and Svitlana Claudia Hluvko. Florence: Edifir, 2002.

Connor, Carolyn L. *Saints and Spectacle: Byzantine Mosaics in Their Cultural Setting*. New York: Oxford University Press, 2016.

Dalton, O. M. "The Gilded Glasses of the Catacombs." *Archaeological Journal* 58 (1901): 225–53.

Federici-Vescovini, Graziella. *Le teorie della luce e della visione ottica dal IX al XV secolo: Studi sulla prospettiva medievale e altri saggi*. Perugia: Morlacchi Editore, 2003.

Gardner, Julian. "Saint Louis of Toulouse, Robert of Anjou and Simone Martini." *Zeitschrift für Kunstgeschichte* 39 (1976): 12–13.

Grant, Edward, ed. *A Source Book in Medieval Science*. Cambridge, MA: Harvard University Press, 1974.

Hills, Paul. *The Light of Early Italian Painting*. New Haven, CT: Yale University Press, 1987.

Hutton, Edward. *The Cosmati: The Roman Marble Workers of the Twelfth and Thirteenth Centuries*. London: Routledge and Paul, 1950.

Kanter, Laurence B., Barbara D. Boehm, Carl B. Strehlke, Gaudenz Freuler, Christa C. Thurman and Pia Palladino. *Painting and Illumination in Early Renaissance Florence, 1300–1450*. New York: Harry N. Abrams, 1994.

Leonards, Ute, Roland Baddeley, Iain D. Gilchrist, Tom Troscianko, Patrick Ledda, and Beth Williamson. "Mediaeval Artists: Masters in Directing the Observers' Gaze," *Current Biology* 17, no. 1 (January 2007): R8–R9.

Lindberg, David. *Theories of Vision from Al-Kindi to Kepler*. Chicago: University of Chicago Press, 1976.

Livingstone, Margaret. *Vision and Art: The Biology of Seeing*. New York: Abrams, 2002.

Macfarlane, Alan, and Gerry Martin. *Glass: A World History*. Chicago: University of Chicago Press, 2002.

Maginnis, Hayden. "Assisi Revisited: Notes on Recent Observations." *The Burlington Magazine* 117, no. 869 (August 1975): 511–17.

Milanesi, Carlo and Gaetano, eds. *Il libro dell'arte o Trattato della pittura, di Cennino Cennini da Colle di Valdelsa; di nuovo pubblicato con molte correzioni e coll'aggiunta di più capitoli tratti dai codici fiorentini, per cura di Gaetano e Carlo Milanesi*. Florence: Felice Le Monnier, 1859.

Muth, Claudia and Claus-Christian Carbon. "The Aesthetic Aha: On the Pleasure of Having Insights into Gestalt." *Acta Psychologica* 144, no. 1 (September 2013): 25–30.

Nelson, Robert S., ed. *Visuality Before and Beyond the Renaissance: Seeing as Others Saw*. Cambridge: Cambridge University Press, 2000.

"Origin and Influence, Cultural Contacts: Egypt, the Ancient Near East, and the Classical World." *The Metropolitan Museum of Art Bulletin* 29, no. 7 (March 1971): 318–26.

Oryshkevich, Irina Taïssa. "The History of the Roman Catacombs from the Age of Constantine to the Renaissance." Ph.D. dissertation, Columbia University, 2003.

Pajares-Ayuela, Paloma. *Cosmatesque Ornament: Flat Polychrome Geometric Patterns in Architecture*. New York: Norton, 2001.

Panofsky, Erwin. *Abbot Suger on the Abbey Church of St. Denis and Its Art Treasures*. Princeton, NJ: Princeton University Press, 1979.

Pettenati, Silvana. *I vetri dorati graffiti e i vetri dipinti*. Turin: Museo Civico di Torino, 1978.

———. "The Decorated Glass." In *Giotto: The Santa Maria Novella Crucifix*, translated by Mark Roberts and Svitlana Claudia Hluvko, 203–15. Florence: Edifir, 2002.

Price, Jennifer. "Glass." In *A Handbook of Roman Art*, edited by Martin Henig, 205–19. London and New York: Phaidon Press, 2006.

Rossi, Ferdinando. *Mosaics: A Survey of Their History and Techniques.* New York: Praeger, 1970.

Schöne, Wolfgang. *Über das Licht in der Malerei [About the Light in Painting].* Berlin: Mann, 1954.

Schwarzenberg, Erkinger. "Colour, Light, and Transparency in the Greek World." In *Medieval Mosaics: Light, Color, Materials*, edited by Eve Borsook, Fiorella Gioffredi Superbi, and Giovanni Pagliarulo, 15–34. Florence: Silvana, 2000.

Sedlmayr, Hans. *Über Farbe, Licht und Dunkel* [On colour, Light and Darkness]. München: Hueber, 1959.

Shelton, Lois Hiedmann. "Gold in Altarpieces of the Early Italian Renaissance: A Theological and Art Historical Analysis of its Meaning and of the Reasons for its Disappearance." Ph.D. dissertation, Yale University, 1987.

Squadrani, Irenaeus. "Tractatus de luce Fr. Bartholomaei de Bononia." *Antonianum* 7 (1932): 201–38.

Tait, Hugh, ed. *Five Thousand Years of Glass.* London: British Museum Press, 1991.

Thompson, Nancy. "The Franciscans and Stained Glass in Tuscany and Umbria." In *Mendicant Cultures in the Medieval and Early Modern World: Word, Deed, and Image*, edited by Sally J. Cornelison, 22–44. Turnhout: Brepols, 2015.

Thunø, Erik. "Inscription and Divine Presence: Golden Letters in the Early Medieval Apse Mosaic." *Word and Image* 27, no. 3 (2011): 279–91.

White, John. *Art and Architecture in Italy: 1250–1400.* New Haven, CT: Yale University Press, 1993.

Zajonc, Arthur. *Catching the Light: The Entwined History of Light and Mind.* Oxford: Oxford University Press, 1993.

4

Transparent Glass from the East

Beruni, Hunain, and Alhazen

As described in the previous chapters, artists ranging from Nicola Pisano to Simone Martini to Paolo di Giovanni Fei incorporated stained glass and gilded glass into their sculptures, frescoes, and panel paintings in order to create striking visual effects that resonated with late medieval and early Renaissance religious symbolism and visuality. This chapter and the next continue this investigation but focus on yet another type of glass, namely transparent glass. Analyzing how early Renaissance artists used transparent glass reveals important insights into the relationship between art, technology, and visuality. But, before doing so, it is necessary to consider the influence of eastern glass, that is, glass made in Byzantine and Islamic territories located on and around the eastern coast of the Mediterranean Sea.

Eastern traditions and practices profoundly impacted glass production in the Latin West. Not only was glass originally invented in the Levant but, after the fall of the Roman Empire when European glassmaking decreased in production, the glass industry in the East flourished in terms of both activity and inventiveness, keeping the traditional glassmaking practices alive. During the late medieval period eastern glassmakers were largely responsible for reinvigorating Europe's glass industry. In addition to exporting their techniques and materials, glassworkers from the Levant also supplied the West with a variety of impressive glass vessels, many of which made their way to Europe as trade goods, loot, diplomatic gifts, and souvenirs from pilgrimages to the Holy Land.

Venice in particular benefitted from its extensive trade and diplomatic relations with Islamic territories.[1] The Serene Republic had access to exquisite glass luxury products, the materials needed to make glass, and glassmakers from the Levant.[2] Raw materials such as ash and alkali from Syria provided Venice with the ingredients needed to produce high quality glass from scratch while imported cullet, that is broken fragments of glass, offered a more economic option. These shards of glass could be melted at a much lower temperature and thus efficiently recycled into new products at a much lower cost.[3]

By at least the thirteenth century Venetians had also mastered the secret for making transparent glass. Archaeological finds and documentary evidence supports this; there are fragments of clear glass and a treatise describing the process for decolorizing glass which date to this time.[4] Suggesting that the colorless glass was both widely available and relatively high in quality is a decree from April of 1300 that banned imitation rock crystal. Few, if any, other materials could rival the transparency of crystal and thus the presumption is that the colorless glass produced in Venice during the early fourteenth century was so clear that it could be mistaken for—or compete with—rock crystal.[5]

Islamic influence on Venetian glass is clearly evident when comparing the fourteenth-century Venetian *Aldrevandin Beaker* in the British Museum (inventory no. 1876,1104.3) to Syrian prototypes such as the *Beaker* from the Walters Art Gallery from ca. 1260 (Figure 4.1).[6] Both vessels are made of transparent glass, feature enameled decoration, and slightly taper toward the bottom. Measuring about five and six-and-a-half inches tall respectively, the sloping contours of these vessels would have conformed nicely within one's grasped hand during use. Similar glasses appear in several important fourteenth-century paintings attesting to their popularity. For instance, Giotto's panel of the *Last Supper* in Munich's Alte Pinakothek (inventory no. 643) and Duccio's panel of the same subject for his *Maestà* for Siena's cathedral (Figure 2.3) both feature depictions of these glass vessels, indicating that they were circulating throughout the Italian market by the first decade of the fourteenth century.

Glassware like the Syrian beaker could have been imported or obtained in the Levant as a souvenir from one's visit to the *loca sancta*. Artworks such as these, which were produced in lands occupied by the crusaders and thus known as Crusader art, blend Islamic techniques, styles, and even calligraphy with subject matter that was appealing to a Christian audience.[7] In this case the glass features an Arabic inscription and a scene of a figure riding a donkey, a subject

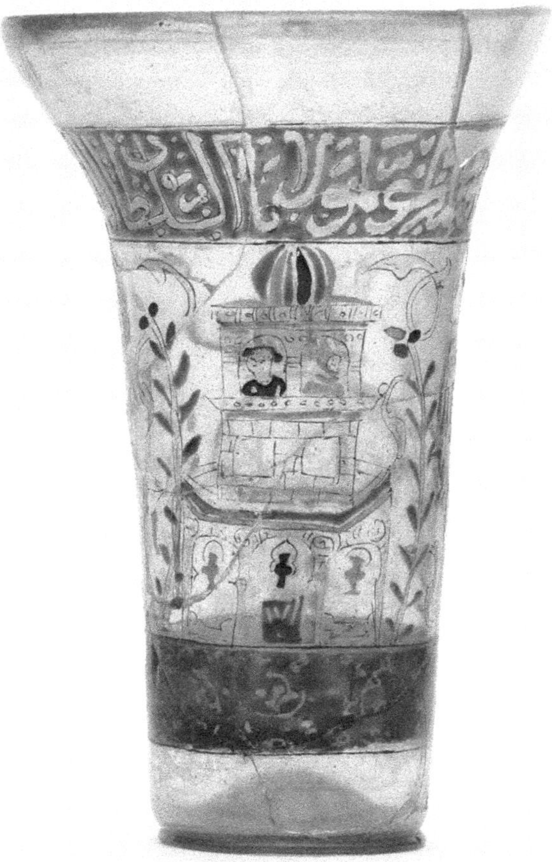

Figure 4.1: Syrian Workshop, *Beaker*, ca. 1260, The Walters Art Museum, Baltimore (47.17). Source: Public Domain, The Walters Museum (CC0).

that could have easily resonated with the story of Christ's entry into Jerusalem. This type of contact and exchange between the Latin West and the Islamic East, which was fostered by the crusades and various economic interests, shaped fourteenth-century Italian visual culture by introducing new types of glass products and techniques for making glass and contributing to the revival of the Italian glass industry.

Glass Reliquaries from the Holy Land

Pilgrim flasks from the Holy Land were some of the most prized glass objects to enter the Latin West, not necessarily for their material value, but rather for their sacred contents.[8] Despite changing political fortunes and increased expense and danger, Christians continually made pilgrimages to the Holy Land from the fourth century onwards.[9] Spiritual tourists to the *loca sancta* were interested in making physical contact with the Biblical landscape and, if possible, bringing back a tangible reminder of this experience, a practice that was well established at least by the sixth century.[10] Christians venerated the remains of saints throughout the medieval and Renaissance periods and found them a powerful source of devotional inspiration. The most valuable object a Christian could encounter was a primary relic, that is, a fragment of a holy person's body. Because primary relics were such exceptional cases they were oftentimes controlled by their elite owners or highly regulated by church officials and thus not readily accessible to the average pilgrim. The same was true for secondary relics, objects owned by a saint or objects that had made extensive physical contact with a saint. Much more available to the average pilgrim were tertiary relics, objects or liquids that had been placed in close proximity to a relic thereby obtaining some of the relic's virtue. These sometimes took the form of *brandea* (strips of cloth temporarily placed near a relic) or *eulogia* (term meaning "blessing" and used to describe other substances placed near relics).

Some of the most coveted *eulogia* included oil from the Holy Sepulcher; balsam from Matarieh, which marked the spot where the Holy Family had spent time on their flight into Egypt; oil-like "tears" cried by an icon of St. Mary at Serdinale; oil from lamps in sacred settings; wax from candles in significant shrines or locations; and water and soil from holy sites.[11] Another type was oil that was imbued with the sanctity of relics by way of contact with primary relics. At many sites throughout the Holy Land liquid substances such as oil were used to anoint holy remains and, after the ritual, the oil was gathered into small vessels and distributed to pilgrims. Or in other cases vessels prefilled with oil were placed near the primary relic during the ceremony and given to visitors afterwards.[12] These containers could be made from a variety of materials such as terracotta or metal but those of greatest interest to this study were the ones made from glass.

One of the earliest documented records of a glass reliquary obtained in the Holy Land and brought to the West comes from the account of the Franciscan friar Salimbene, who records how his Franciscan brother served as a missionary in Egypt during the second crusade of King Louis IX in 1270 and "brought home Manna in a vessel of glass" along with other important relics such as water from the well at Matarieh and balsam wood.[13] Salimbene does not specify the material of the containers used to store the water or wood but he does specifically describe that the vessel containing the manna was made from glass. Although he does not elaborate on the appearance of glass vessel, it is tempting to think that Salimbene mentioned the glass object because it caught his attention for some reason. Perhaps it was because, as will be argued later in this chapter, glass—especially transparent or highly translucent glass—was ripe with potential Christian symbolism by the late thirteenth century and throughout the fourteenth century and its use as a vehicle for *eulogia* made it all the more captivating.

A later pilgrimage account from ca. 1350 provides a description of the distribution of glass jars at a holy site. When making a visit to the holy image of the Virgin near Damascus an anonymous pilgrim records how the sacred painting was "entirely converted into a fleshy substance, so that it ceases not night and day to emit a sacred oil, which the pilgrims who come there from every quarter carry away in little glass jars."[14] This account is especially informative because the pilgrim does not mention a variety of different types of vessels available at the site. Rather his comment suggests that the flasks were made exclusively, or at least primarily, from glass. Such a situation is not surprising considering that Syria was well known for its glass production and therefore glass would have been readily available in this region. It is also possible that Christian pilgrims made connections between their glass reliquaries and the local glass industry, a theory supported by another pilgrimage account by Lionardo Frescobaldi.

During his trip to Hebron in 1382 Frescobaldi recorded that he observed many people gathering sacred oil from the tomb of Abraham, Isaac, and Jacob. As he writes, "From the grave of the said patriarch a certain oil is got, in which the Saracens, the Jews and all the Christians of these parts, as we also, have great faith, and some of us got some of it."[15] Though he does not specify the material of the jars, immediately after making this comment he notes that Hebron "is a very beautiful city and a beautiful country, and very industrious and they do there the finest work in glass, and more than in any place I have been."[16] Thus it is possible that some, if not most, of the containers here were made from glass. That Frescobaldi connects his experience obtaining relics to the area's glass production suggests that this pilgrim, and perhaps others, may have been conscious of the correlation between the objects and their method of production, a theory considered in more detail below.

The large number of extant glass pilgrim flasks of eastern origin now found in western collections attests to their popularity, especially considering the decreased chance of survival of such fragile objects. Generally speaking two main types of vessels were used as pilgrim reliquaries: flasks and bottles. Pilgrim flasks, called *ampullae*, are usually small, somewhat flat, rounded vessels with handles connecting the neck to the body. Fine examples of glass *ampullae* can be found in the Cleveland Museum of Art (inventory no. 1999.235), Smith College Museum (inventory no. 1954:68–97, 1954:68–98, 1954:68–17), and Brooklyn Museum (Figure 4.2). While many surviving examples are also made from clay and metal, this study will focus on those made of transparent and relatively colorless glass because, as I argue, these objects had certain advantages over other materials.

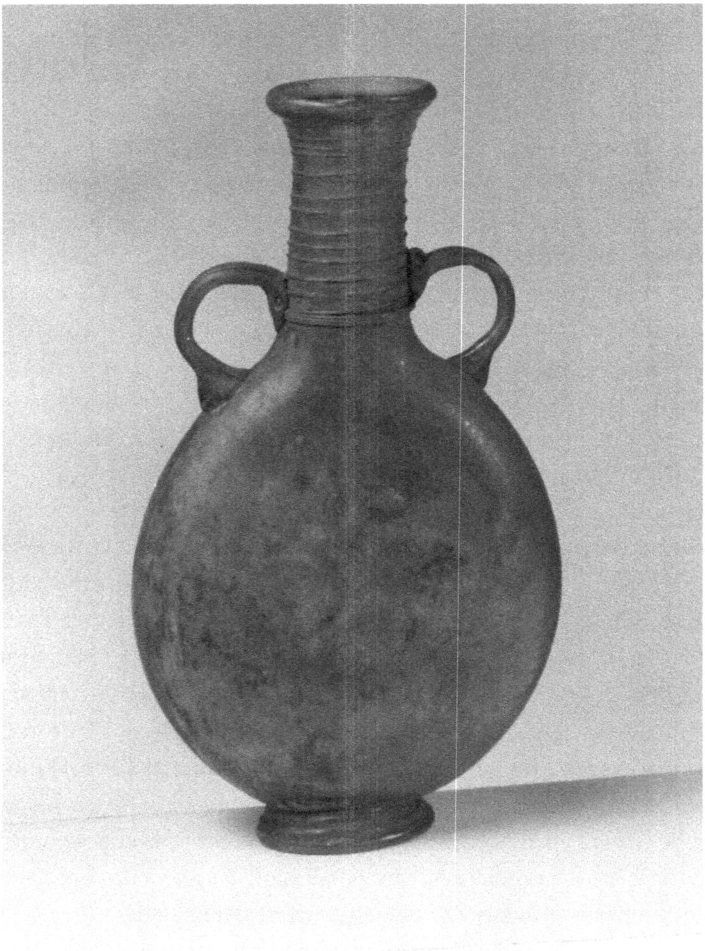

Figure 4.2: Roman Workshop, *Pilgrim Flask*, 1st–5th century, Brooklyn Museum of Art, Gift of the executors of the Estate of Colonel Michael Friedsam (32.739). Source: Brooklyn Museum (CC BY).

Examples of the second category of container, typically cataloged as a bottle or jug, are found in museum collections throughout the world with particularly good examples found in the Metropolitan Museum of Art (Figure 4.3).[17] Dan Barag's research on this type analyzed the Christian and Jewish iconography featured on the bottles, established their method of production, and described their function as reliquaries.[18] The vessels in Barag's group were all made with similar molds and workshop practices and typically feature hexagonal bases, long necks, tubular handles, and widening lips.[19]

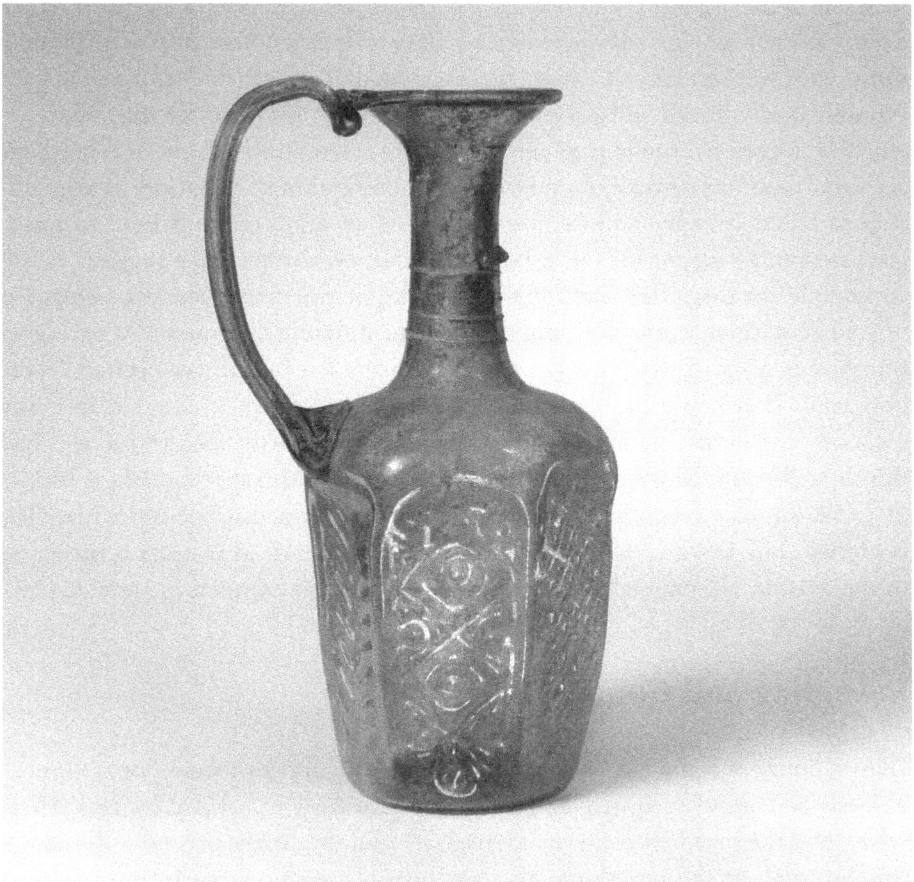

Figure 4.3: Roman-Syrian Workshop, *Glass Hexagonal Jug*, 6th–early 7th century, The Metropolitan Museum of Art, New York (X.243). Source: Public Domain, The Metropolitan Museum of Art (CC0).

Other, more lavish examples of glass reliquaries brought from the Holy Land into the Latin West include a group studied by Avinoam Shalem that consists of four clear glass vessels used as reliquaries or as decoration for reliquaries.[20] The first beaker, now in the treasury of the Cathedral of St. Paulus in Münster, was incorporated into a turriform reliquary during the thirteenth century while the second example, now in the Rheinisches Landesmuseum in Bonn, was treated in a similar manner sometime in the following century.[21] Shalem's other two case studies, enameled bottles now in the Cathedral of St. Stephen in Vienna, were thought to have arrived in Vienna in 1363 containing earth stained with blood from the massacre of the innocents and may have been carried by Rudolf IV upon his return from Constantinople.[22] Though not all glass reliquaries were displayed in such sumptuous ways as these, if a vessel had been used to transport relics it would have certainly been imbued with sacred associations by virtue of contact with relics.

The numerous extant glass reliquaries and references to them in pilgrimage accounts raise important questions about the reception of glass and its relationship to relics. It seems counterintuitive to use such a fragile medium to transport something as precious as a holy relic and yet the evidence suggests it was commonly the case. This leads one to consider that perhaps these glass *ampullae* offered something even more important than durability. Transparent or highly translucent glass vessels provided a visual affirmation of and connection to the holy relic. They could simultaneously reveal and protect their contents in a way most other materials could not. As will be discussed shortly, rock crystal also had this capability but glass may have been better suited to this specific task, as Isidore of Seville suggests when he observes, "Anything contained inside other materials is hidden, but any sort of liquid or visible thing contained in glass is displayed to the outside; although closed up, in a certain way the contents are revealed."[23]

Revealing and Concealing

Glass reliquaries could have resonated with the early modern viewer on a number of levels because of their dual ability to simultaneously reveal and conceal. These vessels had the paradoxical ability to display—and thus make present—the sacred contents and, at the same time, contain them—creating a barrier and denying access to them. The glass vessel's dynamic capability eloquently reflected the nature of the relic inside. Relics embodied both a spiritual power and an earthly presence. Relics were both part of the human world because they were closely connected to the saint's earthly life and they also had a divine component because their ultimate reference point was the saint's heavenly soul. Thus for medieval Christians, relics occupied a liminal space between the sacred and mundane and a close encounter with them could bring the viewer to the threshold of the divine sphere. Thus it should be fitting that their repositories reference this duality.

The visual connection established between the viewer and the holy relic reinforced one's experience of Christ and the *loca sancta* in a powerful way, both along the dangerous journey home and after they arrived. Viewing the relic could have reminded the viewer of their experiences in the Holy Land tracing the steps of the apostles and seeing the sacred sites. Such a visceral connection to the land where the biblical stories took place could have certainly strengthened one's familiarity with the concept of Christ's humanity and, in turn, one's connection to God.

Transparent glass reliquaries could have also evoked the complex nature of the relationship between viewer and God. Like the relics God was both present in the Christian's life and yet not entirely accessible. The early modern Christian might find signs of divinity within the earthly realm but God himself was not fully knowable until after the Beatific Vision, the spiritual union of believer and divine that happened after one's death.

It is very likely that glass held additional symbolic meanings considering the many references made to this medium during the medieval period, particularly those related to the Virgin Mary. As early as the seventh century Venantius Fortunatus described Mary as a church with windows and the light filtering through those windows as God. This notion evolved into one of the Latin West's most famous metaphors involving glass, namely, that the miraculous virgin birth of Christ was like light passing through glass without breaking it.[24] A thirteenth-century verse expresses the concept saying

> As the sunbeam through the glass
> Passeth but not staineth,
> So the Virgin as she was
> Virgin still remaineth.[25]

Yrjö Hirn's seminal book *The Sacred Shrine: A Study of the Poetry and Art of the Catholic Church* observes that "From the beginning of the ninth century theologians, in writing about the virgin birth, commenced to quote the analogy of the passage of light through glass, and poets knew well how to make use of so apt and poetical a simile."[26]

The doctrine which stated that the Virgin Mary was the mother of Christ and yet still remained a virgin was a complicated one but one well worth investigating because it lay at the heart of the God-Christ relationship. Central to this mystery is the notion of the undisturbed passage. The process of viewing a relic contained in a transparent or translucent glass vessel could have eloquently resonated with the symbolism of the virgin birth as well as the other important miracles involving undisturbed passage associated with Christ's death and Resurrection. After Christ's Crucifixion and interment his resurrected spirit passed through the large stone sealing his tomb and, following this, he appeared to his apostles by passing through a closed door. The notion of passage without disturbance—a feat equally important and elusive—was also associated with other important moments in Christ's life.[27] This notion is referenced at both Christ's Incarnation and Resurrection, that is, moments involving a transition between corporeal and divine states of being. For instance when Christ appears to Mary Magdalene and his apostles after the Resurrection, his former companions are able to see but not touch him. That the metaphor of glass was extensively employed to describe the ultimate example of the undisturbed passage, the virgin birth, suggests it may have had a unique place among materials when it came to symbolizing such paradoxical beliefs.

There were many other additional reasons a viewer might have imbued the medium of glass with miraculous or divine associations. The Bible mentions glass when describing the Heavenly Jerusalem, giving this material a place in the celestial sphere. As described in Revelations 21:18, "The wall is built of jasper, while the city is pure gold, clear as glass." Further reinforcing this is Revelations 21:21, which reads, "And the twelve gates are twelve pearls, each of the gates is a single pearl, and the street of the city is pure gold, transparent as glass." Glass was also associated with divine miracles. When describing the miraculously uncorrupted internal organs of deceased Saints Hugh of Lincoln and Edward the Confessor, accounts indicate that the former's organs were "purer than glass" while the latter's flesh was "fair and fresh of colour, pure, and brighter than glass."[28] Glass also played a central role in two miracles recorded by Saint Gregory in his *Life of Saint Benedict*. In the first instance, several monks in disagreement with Benedict's strict rules decided to poison his wine. But before Benedict tasted the tainted drink, he made the sign of the cross. At this point the glass shattered into pieces thereby saving the monk. In another instance, Benedict threw a glass vessel with oil out the window to punish a monk who refused to give the remaining oil to a poor man. But it did not break. Instead a barrel's worth of oil miraculously appeared after the incident.[29]

Thus glass held important symbolic associations with some of the faith's most important mysteries and these associations, in addition to the medium's ability to simultaneously reveal and conceal its contents, made it particularly well suited to function as a reliquary. As pilgrims carried back their sacred souvenirs from the Holy Land the glass would have provided ample meditative avenues to ponder.

Rock Crystal Reliquaries from the East

Ancient and medieval viewers recognized the similar visual properties shared by transparent glass and rock crystal. Thus, as in the case of glass reliquaries, pilgrim flasks made of rock crystal such as that in the British Museum (inventory no. PE AF.3129) also had the ability to simultaneously reveal and protect the holy relics inside and therefore could have held many symbolic associations similar to those just described. In fact *The Golden Legend*, a series of stories about the saints written during the late Middle Ages, describes how both glass and crystal flasks were used to hold miraculous blood that had issued forth from an image of Jesus Christ, indicating their similar nature and function.[30]

Carved crystal products were not locally produced in the West until the late twelfth century and therefore, like glass, the medium may have been associated with the Holy Land because of its eastern origin.[31] Islamic territories were producing high quality artistic carved crystal by the ninth century under the Fatimids of Egypt.[32] European interest in Fatimid carved crystal was greatly fueled after they obtained many crystal vessels—and glass vessels imitating crystal ones—from the dispersal of the Fatimid treasury in 1061.[33] There were also many pieces of Fatimid crystal in the Byzantine imperial treasury and many of these entered European collections after the 1204 Sack of Constantinople.

Avinoam Shalem's research has dispelled the notion that crystal flasks were initially made as secular perfume bottles and convincingly argued that many were likely meant to house relics from their inception.[34] Some such crystal vessels were imported containing relics from the Holy Land while others were converted into reliquaries after their arrival in the West.[35] Inscriptions offering blessings found on the flasks indicate their sacred nature and may have enhanced their function as a reliquary. For instance, the Victoria and Albert Museum's pendant from ca. 1000 with a mount from ca. 1300 (inventory no. M.110–1966) bears the inscription "Hail Mary, full of grace" in Latin, a holy sentiment reminiscent of Gabriel's Annunciation to Mary that reinforces the aforementioned suggestion that glass was associated with the moment of the Incarnation.

A group of Fatimid crystal vessels augmented by decorative moldings or display settings that emphasized a sacred function made their way into religious contexts in the Latin West. Examples including the *Cross of Nikomedes of Borghorst* from ca. 1050 in Pfarrgemeinde St. Nikomedes, Steinfurt-Borghorst (Figure 4.4), the *Pendant in the Form of a Fish* in the Victoria and Albert Museum (inventory no. M.110–1966), the *Flask Reliquary* in the British Museum (inventory no. FBIs.13), and the Metropolitan Museum of Art's *Reliquary of Mary Magdalene*

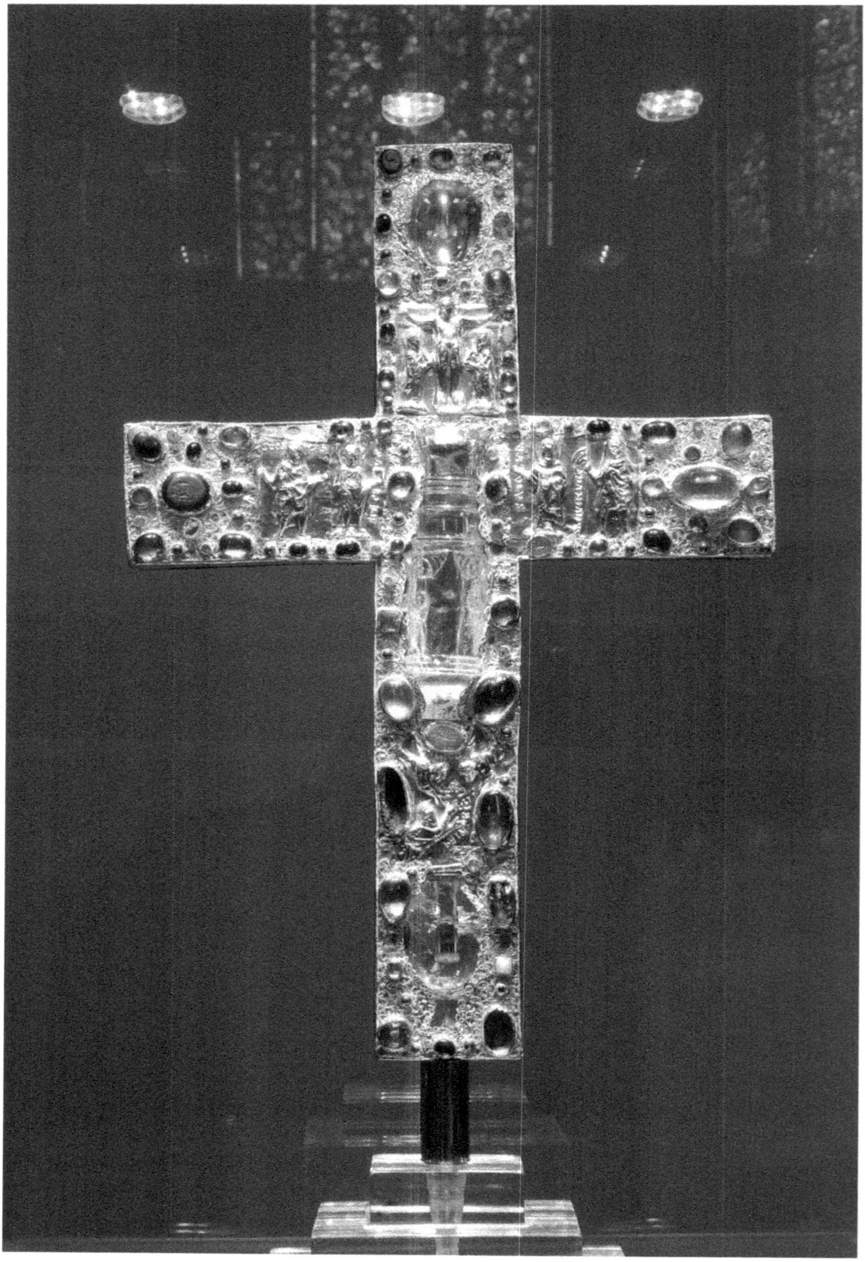

Figure 4.4: German Workshop, *Cross of Nikomedes of Borghorst*, ca. 1050, Pfarrgemeinde St. Nikomedes, Steinfurt-Borghorst, Germany. Source: Markus Cösters via Wikimedia Commons (CC BY 3.0).

(inventory no. 17.190.504) all feature crystal vessels set within a larger devotional artwork. As Shalem points out, when these crystal objects were incorporated into larger devotional ensembles, certain aspects of their design related to their function were lost in a process he calls aestheticization.[36] For example when converting the carved crystal flask into a religious context emphasizing display the weight of the *ampullae* hanging around one's neck or from one's belt could no longer be appreciated. Also inaccessible was the tactile sensation of the faceted angles of the cut crystal. In place of these lost qualities new aesthetic appreciation of the decorations was possible. For instance when the crystal vessel was set within the lavish context of the *Reliquary of Mary Magdalene* or *Cross of St. Nikomedes of Borghorst* the viewer was more likely to appreciate the medium's transparency and the artistry of the engravings in a new, more pronounced way.

Shalem's assessment of these objects' aestheticization is useful and can further be extended to issues related to the objects' materiality. In addition to the newly appreciated aesthetic aspects of the vessels their medium was also now more prominently on display and, subsequently, the visual connection with the relics assumed a more profound role. This new, modified relationship between the viewer and the crystal flask could have easily fostered contemplation of the crystal's transparent qualities, their symbolic potential, and other associations with the medium.

As with glass, ancient sources and biblical references also informed medieval European perception of the medium of crystal. Crystal was associated with natural elements such as air and water. Beruni claimed that crystal was "regarded as noble because of its transparency and clarity, and also because it is like the essential elements of life (i.e., air and water)"[37] and agreed with those, such as Pliny the Elder, who believed that crystal is congealed water. Beruni supports this position by noting how crystal pieces sometimes contain air bubbles, grass, dirt, or other matter which, he reasons, were absorbed by the crystal when it was in its liquid state.[38] Medieval natural scientists in the West, such as Isidore of Seville, disseminated Beruni's theories and extended their influence. As he describes in his *Etymologies*, Isidore posited that crystal was formed from snow that hardened into ice.[39]

The connection between crystal and water fostered the view that crystal had spiritual symbolism related to the sacrament of baptism. For example, Rabanus Maurus described the connection between crystal and the sacrament of baptism in his *De universo*.[40] Thus as with the case of glass vessels, rock crystal ones could have also offered the opportunity to contemplate the visual transmission of the relic through the transparent medium and therefore been a powerful symbolic tool with which to contemplate aspects of the undisturbed passage and the relationship between the heavenly realm and the earthly one.

Beruni on Glass vs. Crystal

Ancient and medieval commentaries also reflect the idea that glass and crystal were related, however, the exact nature of their relationship was debated. Complicating the task of differentiating the two materials is the fact that the Bible and many ancient authors conflate the two substances. For instance, in Revelations 4:5–6 it says, "in front of the throne there is something like a sea of glass, like crystal." Beruni, for his part, clearly distinguishes between glass and crystal yet at the same time, he closely links the two materials. In his early influential account of the ingredients and process for making glass Beruni describes how glass is cast:

> from sand to which borax has been added. The substance is heated for several days on fire till it accumulates, clarifies and progressively hardens. I think—although this is not a virtual certainty—there are different gems in the form of grains in the sand. If you look at it carefully, you will find black, reddish, white and transparent crystalline grains in them.

Beruni's text demonstrates his clear understanding of how to make glass, describing that the process required heating sand and borax until the mixture melted, then cooled, and ultimately hardened. It is his position on the nature of sand, however, that is of more interest here. Upon close examination of the grains of sand, Beruni observed crystalline components. Thus the ingredient needed to make glass was infused with crystal elements, meaning glass contained crystalline elements. In another comment Beruni addressed the relationship between glass and crystal in more detail, noting how some authors distinguish the two based on the fact that glass is artificially produced while crystal occurs naturally. This distinction, though, underscores yet another point of comparison, because as Beruni indicates, the naturally occurring crystal is to be found alongside glass in the very same mines.[41]

There were, however, important concrete differences between glass and crystal. According to Beruni, glass "is softer than crystal and, being more profusely available, is rated lower."[42] This comment reflects the traditional assumption regarding a hierarchy of materials that positions rare, precious, naturally occurring matter more highly than man-made substances, which were seen as imitations. However, as E. Marianne Stern observes, the reputation of glass seems to have evolved over time and may have even been equal to or surpassed the value of gold and lapis lazuli at one point.[43] Thus, while glass was certainly seen as an imitator of more precious gems and crystal at certain times, there were other times when it was highly valued in its own right. Glass indeed had unique properties that distinguished it from crystal and even gave it certain advantages over crystal.

One such advantage related to the material's performance under pressure. As one of Beruni's sources points out "crystal in some ways resembles glass but is not made from it … as regards melting they are distinctly different as, while glass melts, crystal does not."[44] When a crystal object shattered, it was lost forever while, theoretically at least, a glass object could be recast. Beruni recalls many stories about the great loss felt by an owner upon the destruction of a cherished crystal object. He tells the heartbreaking, yet somewhat prudent, story of Alexander the Great who, upon receiving a set of beautiful crystal vessels as a gift, expressed his deep appreciation and then promptly ordered their destruction explaining that he broke them all at once order to escape prolonged sadness and anger directed at his servants who would inevitably break them one by one.[45] Thus, in some sense, glass objects were less perishable because the broken glass could be repurposed. The glass medium's ability to regenerate or reconstitute itself may have only strengthened the aforementioned associations between glass and theological doctrines related to the Incarnation and Resurrection.

Beruni provides commentary on glass's other primary advantage over crystal, noting how "In transparency a glass piece is not very different from the crystal piece, especially if the latter is free from blemishes and bubbles. … Its transparent vessels, which are wholly transparent and show what is insider, are preferred." Beruni here suggests that oftentimes rock crystal contained naturally occurring defects, which impaired its transparency. As crystal was a naturally occurring material one could not control the appearances of blemishes, the best one could do was work around them. As Beruni further explained, "If a hole, knot, or cloudiness tells upon its transparency, it is masked by some etched design or inscription, requiring considerable expertise."[46] These engravings were valued because they required advanced skill by the artist but, it should be noted, that such engravings would have diminished the transparency of the crystal vessel and the visibility of the object inside.

Glass on the other hand could have achieved greater and more consistent transparency relatively easily.[47] If a piece of glass was cast with a defect it could have been melted down and recast until the desired effect was obtained. Thus natural rock crystal may have been more expensive and regarded as more prestigious at various times, but transparent glass had definite advantages when it came to reliquaries and the ability to display the relics. For such reasons, I argue that when it came to making vessels with a consistently high level of transparency at a low cost, glass was better suited to the task.

It is important to consider the potential benefits offered by glass for two main reasons. First, as discussed in the next chapter, fourteenth-century Italian artists began to use transparent glass in a number of important reliquaries. And as will be shown, they were not doing so in imitation of crystal but rather showcasing aspects unique to the glass medium. That is, when fourteenth-century artists in Italy incorporated transparent glass into their reliquaries they did not emulate the appearance or visual effects of rock crystal. Rather the sheets of glass are typically small, flat, round, and thin, very dissimilar from the traditional cabochon shape found with crystals. Rather than mimicking the rounded and polished surfaces of thick rock crystal, these artists embraced formal aspects unique to glass windows.[48] What follows explores the symbolic potential of glass as distinct from that of crystal within the context of optical theory in order to demonstrate that glass could have served a unique role in the veneration of relics.

Hunain, Visual Theory, and Transparent Glass

As mentioned in the introduction chapter it is important to consider artistic glass alongside the medium's other uses, particularly its use in optical devices. The rationale for associating glass and optics begins at the etymological level—the ancient term for glass, *vitrum,* derived from the verb *videre*, meaning "to see."[49] Isidore of Seville confirmed that this connection was known at least into the early medieval era when he noted, "Glass (vitrum) is so called because with its transparency it transmits light to one's sight (visus)."[50]

Perhaps even more significant for this discussion is the fact that at the same time artists were using transparent glass to decorate reliquaries glass was also being used for eyeglasses, which were becoming increasingly popular throughout Italy. As discussed in greater detail in the next chapter, eyeglasses were developed around 1280 in Pisa, with knowledge of them spreading to Florence by at least 1305. One can only assume that their production increased steadily because there are paintings depicting figures wearing glasses made throughout the fourteenth century and documentary evidence of large orders for spectacles being placed by the fifteenth century.[51]

As with the case of glass objects, glassmaking technologies, and raw materials, the Latin West received much of its knowledge about optics and visual theory from eastern sources. While the ancient Greeks had laid the foundation for Europe's understanding of optics, the medieval Christians did not inherent their information directly from classical authors. Rather Arabic translations and Muslim scientists (or non-Muslim, Arabic individuals living under Islamic rule) played a crucial part in the West's revival of learning, especially in transmitting the ideas of Aristotle.[52] The works of Avicenna, Averroes, and Alhazen began filtering into the Latin West in the late eleventh century, with their numbers and influence peaking in the thirteenth century.[53] Islamic efforts to collect and translate ancient learning inspired many original contributions and the impact of these Arabic translations, commentaries, and original optical treatises cannot be overstated.[54]

One of the most striking reasons to consider the connections between visual theory, glass, and the East is the fact that much of the West's optical knowledge was based on an influential ninth-century ophthalmological treatise, *The Book of the Ten Treatises on the Eye,* written by Ḥunayn ibn Isḥāq al-'Ibādī (809–877), known as Hunain or Johannitius in the West. Hunain dramatically shaped the West's understanding of the eye's anatomy and the visual process, transmitting ancient optical knowledge to the West while also adding his own important contributions.[55] Though it was heavily influenced by Galen, Hunain's treatise was more widely available in Europe after 1100 when it, along with many other medical treatises from Arabic and ancient Greek sources, was translated into Latin by Constantinus Africanus.[56]

Most important for the present discussion is the fact that this influential Arabic text describes elements of the eye's anatomy as crystalline and vitreous. As the lens was the seat of vision it occupied the central portion of the eye.[57] Positioned in the middle the lens—or as Hunain describes it, the crystalline humor—which is whitish or uncolored, transparent, luminous, and round with a flattened face, could be easily served by the other elements.[58] The lens's transparency allowed it to quickly receive the colors of perceived objects, its round shape prevented breaking or chipping, and its flatness provided a large surface on which to receive visual stimuli.[59] In his analysis of Hunain's text, Bruce Stansfield Eastwood notes how "The qualities of the lens are those which match the materials it works with in vision—purity of color (white), pure fire or light (luminous), and the clarity of pure air (transparent)."[60] This congruence between the physical properties of the lens and the qualities involved with the visual process result in a successful transmission of visual information. But the lens could not act alone.

The most important supportive membrane of the crystalline lens, which provided it with the necessary nutrients to function, was described by Hunain as the glass-like humor. As he notes, "the vitreous [element] is adjacent to the lens without any partition, and it [the lens] is half submerged in it [the vitreous]."[61]

This vitreous tunic served the crystalline humor by mediating between it and the blood vessels of the retina, a vital function, because:

> Every one of the members of the human body requires nourishment, and this is without doubt because there is a continual loss of its substance ... For this reason it requires a substance to replace that which has been dissolved. But nothing can replace the dissolved substance except that which resembles it, *i.e.* something similar in nature to the member in question. ... A substance is most quickly transmuted into the thing which resembles its own nature most closely. Since the lens without doubt requires nourishment and since, as we mentioned already, this humour is white, transparent and luminous, it is impossible for it to receive its nutrition direct from the blood. It requires an intermediary between its nature and that of the blood; and such is the glass-like humour [the vitreous], as it is nearer to the white colour and transparency than the blood.[62]

Thus the glass-like element's role as intermediary was vital to the visual process. The blood vessels in the retina were too dissimilar from the crystalline lens to transfer nourishment but glass could relate to, or resonate with, both the crystal lens and the bodily fluids in the retina. Because glass was similar to crystal the vitreous humor could act as the supportive membrane to the crystal lens. Although Hunain does not explain how glass could resonate with the blood vessels one can speculate that his reasoning may have stemmed from the fact that glass is has more mundane origins, being made with human hands rather than naturally occurring.

An important theme in Hunain's description of the eye, and throughout the rest of his treatise, is the concept that like affects like. Extending this penchant for corresponding parts to the visual process and the perceived objects, allows one to imagine that glass pilgrim flasks containing relics created a powerful visual encounter because they would extend the symmetry of materials on either side of the point of visual contact. When one envisions the process of the relic's visual stimuli entering the eye, the chain of events would go as follows: the relic's image moves through the glass of the reliquary and then encounters the crystalline lens of the eye. From there, the visual data, moved through the vitreous tunic of the eye, to the blood, and finally onto the mind and heart. There is congruence as glass is the intermediary between the lens and the viewer's body and between the lens and the relic, which represented the saintly flesh and blood. Thus glass is the ideal mediator for both external and internal processes of the visual process. Just as the glass-like component in the eye transmitted nutrients from the body's blood to the lens, the glass reliquaries transmitted a visual image of the relic, the saintly blood and spiritual nourishment, to the seat of vision, the lens.[63] Thus, although the nature of glass and crystal were very similar and both were related to crucial aspects of the eye's anatomy, important differences existed. While it is true that crystal traditionally held a more venerated reputation that glass the latter's distinct qualities may have given it a unique role in the visual process. Glass could perform as a transmitter, or mediator, in a way crystal could not.

There is evidence that the theory of the eye's vitreous nature extended beyond scientific circles and resonated with medieval and Renaissance artists. Artists from a variety of ancient and medieval cultures frequently used glass and crystal insets for eyes.[64] For example, as early as ca. 2500 BCE Egyptian artists were using rock crystal for the eyes of the famous *Seated Scribe* from Saqqara now in the Louvre (inventory no. E 3023). Ancient Greek works also paid special decorative attention to eyes as evidenced by the Metropolitan Museum's *Bronze Statuette of Aphrodite* (inventory no. 35.122) from the late Hellenistic period which features glass paste eyes. In the medieval period, one finds more important examples such as the famous *Reliquary Statue of Sainte Foy*.[65] Finally, as mentioned in the introduction chapter, one of the most public and well-known sculptures of early Renaissance Florence also featured eyes made from glass. Arnolfo di Cambio's *Virgin Mary* for the façade of the Florence cathedral made around 1310 included glass insets for the figure's eyes (Figure 1.1).

As Julian Gardner points out, the glass eyes of Arnolfo's *Madonna* would have been extremely striking when viewed in their original outdoor setting on the façade of the Florence cathedral. The sunlight reflecting off them would have created sharp, prominent, fluctuating—and thus lively—highlights.[66] Such visual effects would have been complemented by the mosaic decoration originally found on the wall behind the Madonna, which would have also sparkled in the sunlight due to glass panels backed with gold. Arnolfo's use of different types of glass is not unexpected from this artist. About four decades earlier he had worked on Nicola Pisano's *Arca,* which was also a tour de force of gilded glass panels. It is not unreasonable to imagine that Arnolfo's experience working on the *Arca* might have introduced him to the medium's powerful visual effects and inspired his later use of glass when designing the sculpture for the cathedral of Florence.

This relationship between the eye and the medium of glass had a long tradition dating as far back as the Bible. As Matthew 6:22 notes, "The eye is the lamp of the body. So, if your eye is healthy, your whole body will be full of light." During the period under discussion here, namely the thirteenth and fourteenth centuries, glass lamps were made in Europe.[67] The scene of the *Verification of the Stigmata* from the Upper Church at San Francesco in Assisi (Figure 4.5) and an illusionistic niche from Giotto's Arena Chapel (Figure 4.6) depict how such small glass lamps were used and displayed.[68]

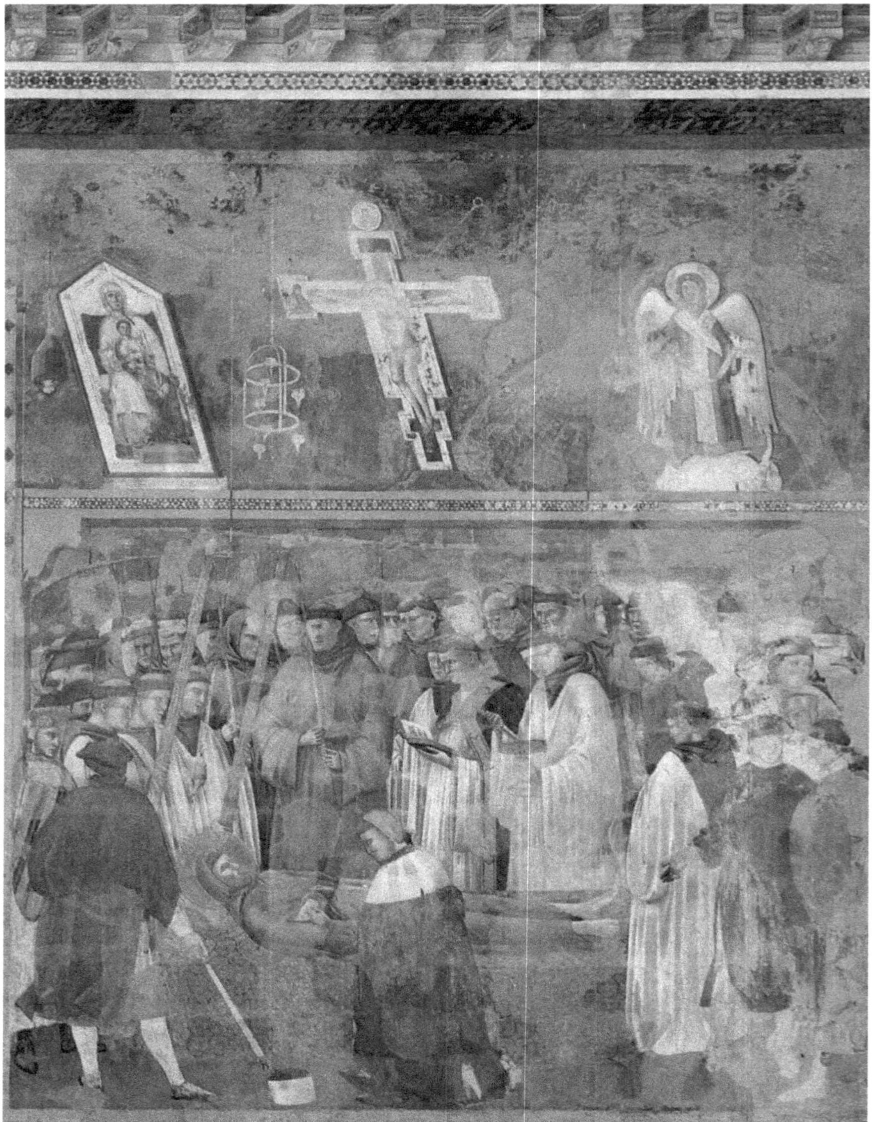

Figure 4.5: Master of the Legend of Saint Francis, *Verification of the Stigmata,* ca. 1300, Upper Church, San Francesco, Assisi. Source: Public Domain via Wikimedia Commons.

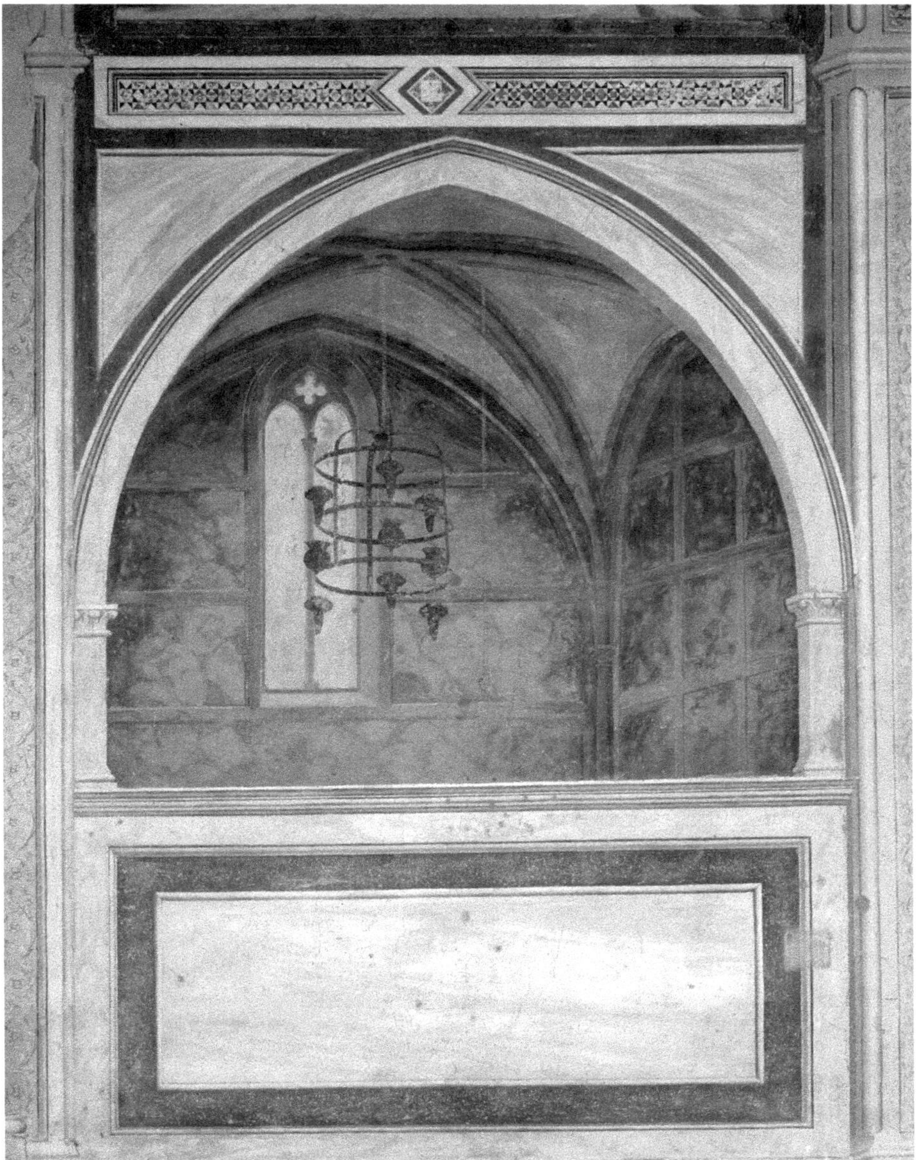

Figure 4.6: Giotto, Detail of an illusionistic niche with hanging lamps, ca. 1305, Arena Chapel, Padua. Source: Raffaello Bencini/Alinari Archives, Florence.

Proverbs, poems, and biblical passages also reference the eye as a window or mirror, both of which were also made from glass during the early modern period. Although most famously credited to Leonardo da Vinci, the notion that the eyes are the windows to the soul was known prior to the fifteenth century.[69] *Cligès* by Chrétien de Troyes, a French poem from around 1176, equates the eye to the mirror of the heart, the eye to a glass lantern, and even incorporates the aforementioned concept of light passing through glass without breaking it.[70] Such references could also have been found within religious texts. Sedulius's *Carmen paschale* from the early fifth century described Christ's miraculous heeling of blindness as "opening the long-closed windows on their countenance"[71] In the seventh century, in his *Life of Saint Martin*, Venantus Fortunatus writes of his experience visiting the tomb of Saint Martin, noting how

> I quickly came closer, in great pain, groaning because light was fleeing from the windows of my eyes. As soon as I touched my eyelids with the consecrated oil, the fiery cloud disappeared from my face, and the physician drove off the malady with his mild unguent.[72]

The thirteenth-century *Le Songe du Castel* echoes this sentiment, likening a man to a castle and his eyes to the windows.[73] Chaucer also makes reference to glass when describing eyes, noting that the prioress's eyes were as grey as glass while the eyes of the miller's daughter are said to be blue as glass.[74]

Conclusions on Transparent Glass from the East: Shaping Renaissance Visuality

The symbolic potential of glass discussed here—both its many religious and optical associations—establishes an important foundation for the next chapter, which conducts a close reading of fourteenth-century Italian reliquaries featuring glass. As the present chapter has shown, in the decades preceding the making of the reliquaries, the Latin West was simultaneously absorbing Arabic optical theory and Islamic glassmaking resources as well as actual glass and crystal vessels from the Levant, many of which functioned as reliquaries. It is for such reasons that one should interpret eastern glass vessels as more than mere spolia or luxury items. Instead it is important to consider how the Latin West's understanding of eastern glass objects was shaped by the spiritual power of the relics, the optical theory and glass traditions of the East, and the act of viewing. As the next chapter demonstrates central Italian artists seem to have understood how to capitalize on these various elements in their attempts to venerate relics in the most powerful way possible and, in so doing, used transparent glass to create something uniquely suited to and characteristic of evolving trecento Italian visuality.

Notes

1. For more on the sustained trade connections despite the conflict of the crusades see Rosamond E. Mack, *Bazaar to Piazza* (Berkeley, CA: University of California Press, 2002), 15. For Venice's relation to the East beginning in the ninth century see Stefano Carboni, *Venice and the Islamic World: 828–1797* (New York: Metropolitan Museum of Art, 2007), 16–17.
2. Mack, *Bazaar to Piazza*, 116, specifically "a Greek from Morea in the Peloponnisos" and "Bartolomeus and his brother Donino from Dalmatia," who recorded there during the late-thirteenth century.
3. The import of cullet peaked around the middle of the thirteenth century. Attesting to the large amount of cullet entering Venice is a treaty from 1277 between Doge Giacomo Contarini and Bohemond VII, prince of Antioch, describing duties placed on cullet from Tripoli. Ibid., 113.
4. Ibid., 114.
5. Ibid.
6. For a more detailed comparison between these objects see ibid., 114–15; John Carswell, "The Baltimore Beakers," in *Gilded and Enamelled Glass from the Middle East*, ed. Rachel Ward (London: British Museum Press, 1998): 61–63; and Marco Verità, "Analyses of Early Enamelled Venetian Glass: A Comparison with Islamic Glass," in ibid., 129–34.
7. Jaroslav Folda, *Crusader Art in the Holy Land: From the Third Crusade to the Fall of Acre, 1187–1291* (Cambridge, NY: Cambridge University Press, 2005), 364.
8. For commentary on non-reliquary eastern glass and its reception in the medieval West see, Avinoam Shalem, "Some Speculations on the Original Cases Made to Contain Enamelled Glass Beakers for Export" in *Gilded and Enamelled Glass from the Middle East*, ed. Rachel Ward (London: British Museum Press, 1998), 64–68; J. M. Rogers, "European Inventories as a Source for the Distribution of Mamluk Enamelled Glass" in ibid., 69–73; and Hugh Tait, "The Palmer Cup and Related Glasses Exported to Europe in the Middle Ages," in ibid., 50–55.
9. Diana Webb, *Medieval European Pilgrimage, c. 700–c. 1500* (Gordonsville, VA: Palgrave Macmillan, 2002), 2–3.
10. Dan Barag, "Glass Pilgrim Vessels from Jerusalem: Part I," *Journal of Glass Studies* 12 (1970): 48.
11. Avinoam Shalem, *Islam Christianized: Islamic Portable Objects in the Medieval Church Treasuries of the Latin West*, 2nd ed. (Frankfurt am Main: Peter Lang, 1998), 18–24.
12. Ibid., 19.

13. G. G. Coulton, *From Saint Francis to Dante: Translations from the Chronicle of the Franciscan Salimbene (1221–1288),* 2nd ed. (Philadelphia: University of Pennsylvania Press, 1972), 172. A more complete version of the quote is as follows: "… and he saw an Unicorn and the Balsam Vine, and brought home Manna in a vessel of glass, and water from Saint Mary's Well (with which alone the Balsam Vine can be watered so as to bear fruit): and Balsam wood he brought home with him, and many such things which we had never seen …"
14. J. H. Bernard, ed., *Guide-Book to Palestine: Circ. A. D. 1350* (London: Palestine Pilgrims' Text Society, 1894), 38. The full quote reads:

> Ten miles from Damascus is the city of Sardinaya, in which is the venerated image of the glorious Virgin Mary, which was brought from Jerusalem. This blessed image was entirely converted into a fleshly substance, so that it ceases not night and day to emit a sacred oil, which the pilgrims who come there from every quarter carry away in little glass jars. No Saracen can live in this city; they always die within a year.

15. Leonardo Frescobaldi, Giorgio Gucci, and Simone Sigoli, *Visit to the Holy Places of Egypt, Sinai, Palestine, and Syria in 1384* (Jerusalem: Franciscan Press, 1948), 68. This source also described the pilgrimage of Niccolo da Poggibonsi, who travelled to the Holy Land around 1346–50.
16. Frescobaldi, *Holy Places,* 68. Niccolo da Poggibonsi also remarks on Hebron's glass industry and that of Damascus, see ibid., 58, 79.
17. A subset of the bottle category includes the vial format, which is similar to the bottle but is usually smaller and cylindrical, is sometimes called a *lacrimarium, ungentarium,* or *balsamarium.*
18. Dan Barag, "Vessels Part I," 48. For more on this, see Dan Barag, "Glass Pilgrim Vessels from Jerusalem: Parts II and III," *Journal of Glass Studies* 13 (1971): 45–63 and John Wilkinson, "Christian Pilgrims in Jerusalem during the Byzantine Period," *Palestine Exploration Quarterly* 108, no. 2 (1976): 84.
19. Although Barag's original study identified brown as the predominant color used for these objects, since its publication, more colorless or relatively colorless examples have come to light.
20. While it's true that these vessels include colorful decoration, the glass is relatively clear and thus the defining feature of transparent glass—that it reveals its contents—is still a valid consideration. Shalem, "Some Speculations," 64–68.
21. Shalem, "Some Speculations," 64.
22. Ibid., 64–65.
23. Stephen A. Barney and others, eds., *The Etymologies of Isidore of Seville* (Cambridge: Cambridge University Press, 2010), 328.
24. Yrjö Hirn, *The Sacred Shrine: A Study of the Poetry and Art of the Catholic Church* (Boston: Beacon Press, 1957), 343–45. For more on this analogy, see Elizabeth

Carson Pastan, "Glazing Medieval Buildings," in *A Companion to Medieval Art*, ed. Conrad Rudolph (Oxford: Blackwell Publishing, 2006), 445; Millard Meiss, "Light as Form and Symbol in Some Fifteenth-Century Paintings," *Art Bulletin* 27, no. 3 (Sept., 1945): 176; and M. B. Ogle, "The Classical Origin and Tradition of Literary Conceits," *The American Journal of Philology* 34 no. 2 (1913): 137. Hirn, *Sacred Shrine*, 343 notes that Saint Bridget described the miraculous conception as God entering "the body of the Virgin just as the sun shines through purest stone or glass." Meiss, "Light as Form," quotes Saint Bernard as follows

> Just as the brilliance of the sun fills and penetrates a glass window without damaging it, and pierces its solid form with imperceptible subtlety, neither hurting it when entering nor destroying it when emerging: thus the word of God, the splendor of the Father, entered the virgin chamber and then came forth from the closed womb.

25. Mrs. Henry Jenner, *Christian Symbolism* (London: Methuen, 1910), 39 and quoted in Peter Fingesten, "Topographical and Anatomical Aspects of the Gothic Cathedral," *The Journal of Aesthetics and Art Criticism* 20, no. 1 (Autumn 1961): 3–23.
26. Hirn, *Sacred Shrine,* 343.
27. The concept of the undisturbed passage was also expressed using the metaphor of the closed door, or *porta clausa.*
28. Adam of Eynsham's account of the preparation of Saint Hugh of Lincoln's body, notes that the saint's internal organs remained unsullied, that they were "purer than glass." See John Crook, *English Medieval Shrines* (Woodbridge, UK: Boydell Press, 2011), 220. For the story of Edward's uncorrupted corpse, see Jacobus de Voragine, *The Golden Legend or Lives of the Saints*, vol. 6, trans. William Caxton, ed. F. S. Ellis, Temple Classics (London: J. M. Dent and Co., 1900), 40.
29. Gregory the Great, "Life of Saint Benedict" in *Early Christian Lives*, trans. Caroline White, Penguin Classics (London: Penguin Books, 1998), 163 & 187.
30. Jacobus de Voragine, *The Golden Legend*, 132.
31. Shalem, *Islam Christianized,* 146. Shalem makes the same point at ibid., 58. For more on the history of carved rock crystal, see Genevra Kornbluth, *Engraved Gems of the Carolingian Empire* (University Park, PA: Pennsylvania State University Press, 1995). Also, Beruni identifies some of the more popular centers of crystal production noting that crystal was

> brought from the island of Zanj and other islands to Basrah, where vessels are made. Large and small pieces are collected at one place. Instructions are tagged upon pieces that are to be cut and shaped and the types of vessels that are to be made from them. They are then handed over to the artisans who follow the instructions and collect high wages. … It is brought from Kashmīr also. Some sections are uncut and some are used in the making of vessels and utensils, goblets and cups, chess pieces and

> counters, and pieces as large as the soap-nut. But this variety does not approach the Zanjī kind nor is the quality of the workmanship of these people (i.e., of Kashmīr) as finished as that of the Basrans. Its sections are found in mountains as well. It is found in plenty in Wakhān and Badakhshān but is not exported. Al-Kindī writes 'The best crystal is the A'rābī which is picked from the desert among the gravel. It is found encapsulated in a thin turbid sheath and weighs up to two ratls. It is similarly picked from Serandīb, but it is less transparent than the A'rābī.'

See Muḥammad ibn Aḥmad Bīrūnī, *The Book Most Comprehensive in Knowledge on Precious Stones: Al-Beruni's Book on Mineralogy,* ed. Hakim Mohammad Said (Islamabad: Pakistan Hijra Council, 1989), 159–60.

32. Though the Fatimid Dynasty is associated with the production of the crystal vessels, Shalem cites documentary evidence that there was an earlier tradition dating back to at least the ninth century. Shalem, *Islam Christianized,* 26.
33. For more on the Fatimid Treasury's dispersion, see Shalem, *Islam Christianized,* 56–66 and for more on the treasury in general, see Helen Romberg, "The Fatimid Treasury: Content and Function" (M.Phil. thesis, University of Oxford, 1985). The *Buckley Ewer* from the Victoria and Albert Museum is a good example of a glass Fatimid object imitating a crystal vessel.
34. Shalem, *Islam Christianized,* 30–31 and Avinoam Shalem, "Islamic Rock Crystal Vessels—Scent or Ampullae?", *Bamberger Symposium: Rezeption in der islamischen Kunst,* Beiruter Texte und Studien 61 (1999): 289–99.
35. Shalem, *Islam Christianized,* 18–31. For more on some small phials listed in fourteenth-century papal inventories which were probably pilgrimage souvenirs, see Mack, *Bazaar to Piazza,* 116.
36. Shalem, *Islam Christianized,* 131.
37. Beruni, *Precious Stones,* 158. At ibid., 157 Beruni goes into more detail noting

> The beryl stone which is called *mahā* and *mihā* is said to have its origin in *mā'* (lustre, water). Because of its limpidity and resemblance to the transparency of water, it began to be called *mahā*. The word, *mā'* is said to have its origin in *mawh,* as its double collective nouns are *miyah* and *amwāh: mawhatu al-shay',* is also derived from the same usage, and is spoken of in connexion with a thing which is endowed with a lustre which it previously did not have. This word is also employed when something is inundated with water or made brilliant or sharp.

38. In *Precious Stones,* 163–64, Beruni writes,

> A person who saw crystal-cutters in Basrah says that they find that crystal pieces have grass, wood, gravel, earth and air-bubbles in them. All this testifies to the fact that the crystal in its pristine state was a liquid in motion. This should not come as a surprise as we find things in certain places that have petrified, and when a plant

or animal can become converted into stone, it is not strange for earth or water to be petrified, and if people did not see these occurrences often, they could not have become known to the commonalty of mankind.

Beruni's source may have been Pliny the Elder, see Pliny the Elder, *The Natural History*, trans. D. E. Eichholz, vol. 10 (Cambridge, MA: Harvard University Press, 1962), 181.

39. Stephen A. Barney and others, *The Etymologies of Isidore of Seville* (Cambridge: Cambridge University Press, 2010), 325 & 328.
40. Robert M. Durling and Ronald L. Martinez, *Time and the Crystal: Studies in Dante's Rime Petrose* (Berkeley: University of California Press, 1990), 253. See also ibid n. 186, which states "In his *De universo* (PL 111.472), Rabanus Maurus, after showing that crystal signifies baptism and the immutability of the angels, concludes: 'Aqua ergo in crystallum versa est quando corruptionis eam infirmitatem, per resurrectionem suam ad incorruptionis firmitatem est immutata.'"
41. Beruni, *Precious Stones*, 160.
42. Ibid., 192.
43. Stern, E. Marianne, "Glass and Rock Crystal: A Multifaceted Relationship," *Journal of Roman Archaeology* 10 (1997): 192–93.
44. Beruni, *Precious Stones,* 160.
45. Ibid., 162.
46. Ibid., 159.
47. Stern, "Glass and Rock Crystal," 194 gives a similar analysis of blue glass and lapis lazuli.
48. This is not to say that crystals could not offer dynamic spiritual viewing. For more on the complex visual effects of rock crystal, see Genevra Kornbluth, "Active Optics: Carolingian Rock Crystal on Medieval Reliquaries," *Different Visions* 4 (2014), 14–22.
49. Vincent Ilardi, *Renaissance Vision from Spectacles to Telescopes* (Philadelphia: American Philosophical Society, 2007), 38.
50. Barney, *Etymologies,* 328.
51. Ilardi, *Renaissance Vision,* 88–100.
52. One of the ways these texts may have been transmitted may include the translation efforts at the court of Alfonso X of Castille. Brunetto Latini was the Florentine ambassador to Alfonso's court from 1259–1260. For more, see Heather Coffey, "Encountering the Body of Muhammad: Intersections between Mi'raj Narratives, the Shaqq al-Sadr, and the Divina Commedia in the Age before Print," in *Constructing the Image of Muhammad in Europe*, ed. Avinoam Shalem (Berlin: Walter de Gruyter, 2013), 41–42.
53. Arthur Hyman and James Walsh, eds., *Philosophy in the Middle Ages* (Indianapolis, IN: Hackett Publishing Company, 1973), 452. Up until about 1240, the main

commentator on Aristotle was Avicenna; from then on, it was Averroes. Alhazen's book on optics was translated into Latin in around 1200 as *On Vision* and was very influential in the period before Kepler according to Alan Macfarlane and Gerry Martin, *Glass: A World History* (Chicago: University of Chicago Press, 2002), 35.

54. David Lindberg, *Theories of Vision from Al-Kindi to Kepler* (Chicago: University of Chicago Press, 1976), 58 asserts "Alhazen was undoubtedly the most significant figure in the history of optics between antiquity and the seventeenth century."
55. Though he was not Muslim he wrote in Arabic and therefore may have been associated with the region of the Holy Land.
56. Lindberg, *Theories of Vision*, 41.
57. Ibid., 34.
58. Ibid., 34–41.
59. Ḥunayn ibn Isḥāq al-ʻIbādī, *The Book of the Ten Treatises on the Eye*, trans. Max Meyerhof (Cairo: Government Press, 1928), 3.
60. Bruce Stansfield Eastwood, "The Elements of Vision: The Micro-Cosmology of Galenic Visual Theory According to Hunayn Ibn Ishaq," *Transactions of the American Philosophical Society* 72, no. 5 (1982): 4.
61. Hunain, *Treatises on the Eye*, 6.
62. Hunain, *Treatises on the Eye*, 6. In full, the passage reads,

> Every one of the members of the human body requires nourishment, and this is without doubt because there is a continual loss of its substance going on through dissolution by reason of the influence of the natural warmth from within and the warmth of the air from without. For this reason it requires a substance to replace that which has been dissolved. But nothing can replace the dissolved substance except that which resembles it, *i.e.* something similar in nature to the member in question. The nutrition is effected in this wise, *viz.* that the member receives an addition of substance resembling its own nature. This accretion, however, can only resemble the nature of the member if the latter transmutes it according to its own nature. A substance is most quickly transmuted into the thing which resembles its own nature most closely. Since the lens without doubt requires nourishment and since, as we mentioned already, this humour is white, transparent and luminous, it is impossible for it to receive its nutrition direct from the blood. It requires an intermediary between its nature and that of the blood; and such is the glass-like humour [the vitreous], as it is nearer to the white colour and transparency than the blood.

63. This interpretation may further strengthen the analysis of Nicola Pisano's *Arca di San Domenico* in Chapter 3. In this case, the gilded glass could have mediated between the saint's blood and the blood fueling the viewer's sense organ.
64. Julian Gardner, "Arnolfo di Cambio: From Rome to Florence," in *Arnolfo's Moment: Acts of an International Conference, Florence, Villa I Tatti, May 26–27, 2005*, eds. David Friedman, Julian Gardner, and Margaret Haines (Florence: L. S. Olschki, 2009), 151 n. 44.

65. Barbara Drake Boehm, "Reliquary Busts: 'A Certain Aristocratic Eminence'," in *Set in Stone: The Face in Medieval Sculpture*, ed. Charles Little (New York: Metropolitan Museum, 2006), 173 n. 24.
66. Gardner, "Arnolfo di Cambio," 151.
67. David Whitehouse, *Medieval Glass for Popes, Princes, and Peasants* (Corning, NY: Corning Museum of Glass, 2010), 53–55 and Daniela Stiaffini, *Il vetro nel medioevo: Tecniche, strutture, manufatti* (Rome: Fratelli Palombi, 1999), 119.
68. Marina Uboldi, "Diffusione delle lampade vitree in età tardoantica e altomedievale e spunti per una tipologia," *Archeologia Medievale* 22 (1995): 93–145, specifically p. 95 for commentary on Giotto's lamps in the Arena Chapel.
69. Martin Kemp, *Leonardo on Painting* (Yale: Yale University Press, 2001), 20.
70. Chrétien de Troyes, *Cligés*, trans. Burton Raffel (New Haven, CT: Yale University Press, 1997), 24–25.
71. Michael John Roberts, *The Humblest Sparrow: The Poetry of Venantius Fortunatus* (Ann Arbor, MI: University of Michigan Press, 2009), 183.
72. Robert Levine, "Patronage and Erotic Rhetoric in the Sixth Century: The Case of Venantius Fortunatus," in *Words of Love and Love of Words in the Middle Ages and the Renaissance*, Medieval and Renaissance Texts and Studies, ed. Albrecht Classen, no. 347 (Tempe, AZ: Arizona Center for Medieval and Renaissance Studies, 2008), 88.
73. Roberta Cornelius, "The Figurative Castle, a Study in the Medieval Allegory of the Edifice with Especial Reference to Religious Writings" (Ph.D. diss., Bryn Mawr College, 1930), 16–17.
74. Geoffrey Chaucer, *The Canterbury Tales*, trans. David Wright (Oxford: Oxford University Press, 2008), 101.

References

Barag, Dan. "Glass Pilgrim Vessels from Jerusalem: Part I." *Journal of Glass Studies* 12 (1970): 35–63.

———. "Glass Pilgrim Vessels from Jerusalem: Parts II and III." *Journal of Glass Studies* 13 (1971): 45–63.

Barney, Stephen A., W. J Lewis, J. A. Beach, and Oliver Berghof, eds. *The Etymologies of Isidore of Seville.* Cambridge: Cambridge University Press, 2010.

Bernard, J. H., ed. *Guide-Book to Palestine: Circ. A.D. 1350.* London: Palestine Pilgrims' Text Society, 1894.

Bīrūnī, Muḥammad ibn Aḥmad. *The Book Most Comprehensive in Knowledge on Precious Stones: Al-Beruni's Book on Mineralogy.* Edited by Hakim Mohammad Said. Islamabad: Pakistan Hijra Council, 1989.

Boehm, Barbara Drake. "Reliquary Busts: 'A Certain Aristocratic Eminence'." In *Set in Stone: The Face in Medieval Sculpture*, edited by Charles T. Little, 168–73. New York: Metropolitan Museum of Art, 2006.

Carboni, Stefano. *Venice and the Islamic World: 828–1797.* New York: Metropolitan Museum of Art, 2007.

Carswell, John. “The Baltimore Beakers.” In *Gilded and Enamelled Glass from the Middle East*, edited by Rachel Ward, 61–63. London: British Museum Press, 1998.

Chaucer, Geoffrey. *The Canterbury Tales.* Translated by David Wright. Oxford: Oxford University Press, 2008.

Coffey, Heather. “Encountering the Body of Muhammad: Intersections between Mi‘raj Narratives, the Shaqq al-Sadr, and the Divina Commedia in the Age before Print.” In *Constructing the Image of Muhammad in Europe*, edited by Avinoam Shalem, 33–86. Berlin: Walter de Gruyter, 2013.

Cornelius, Roberta. “The Figurative Castle, a Study in the Medieval Allegory of the Edifice with Especial Reference to Religious Writings.” Ph.D. dissertation, Bryn Mawr College, 1930.

Coulton, G. G. *From St. Francis to Dante: Translations from the Chronicle of the Franciscan Salimbene (1221–1288).* 2nd ed. Philadelphia: University of Pennsylvania Press, 1972.

Crook, John. *English Medieval Shrines.* Woodbridge, UK: Boydell Press, 2011.

Durling, Robert M. and Ronald L. Martinez. *Time and the Crystal: Studies in Dante's Rime Petrose.* Berkeley: University of California Press, 1990.

Eastwood, Bruce Stansfield. “The Elements of Vision: The Micro-Cosmology of Galenic Visual Theory According to Hunayn Ibn Ishaq.” *Transactions of the American Philosophical Society* 72, no. 5 (1982): 1–59.

Fingesten, Peter. “Topographical and Anatomical Aspects of the Gothic Cathedral,” *The Journal of Aesthetics and Art Criticism* 20, no. 1 (Autumn 1961): 3–23.

Folda, Jaroslav. *Crusader Art in the Holy Land: From the Third Crusade to the Fall of Acre, 1187–1291.* Cambridge, NY: Cambridge University Press, 2005.

Frescobaldi, Leonardo, Giorgio Gucci, and Simone Sigoli. *Visit to the Holy Places of Egypt, Sinai, Palestine, and Syria in 1384.* Jerusalem: Franciscan Press, 1948.

Gardner, Julian. “Arnolfo di Cambio: From Rome to Florence.” In *Arnolfo's Moment: Acts of an International Conference, Florence, Villa I Tatti, May 26–27, 2005*, edited by David Friedman, Julian Gardner, and Margaret Haines, 141–60. Florence: L. S. Olschki, 2009.

Gregory the Great. “Life of St. Benedict.” In *Early Christian Lives.* Translated by Caroline White, 161–204. London: Penguin Books, 1998.

Hirn, Yrjö. *The Sacred Shrine: A Study of the Poetry and Art of the Catholic Church.* Boston: Beacon Press, 1957.

Hyman, Arthur, and James Walsh, eds. *Philosophy in the Middle Ages.* Indianapolis, IN: Hackett Publishing Company, 1973.

Ilardi, Vincent. *Renaissance Vision from Spectacles to Telescopes.* Philadelphia: American Philosophical Society, 2007.

al-ʻIbādī, Ḥunayn ibn Isḥāq. *The Book of the Ten Treatises on the Eye.* Translated by Max Meyerhof. Cairo: Government Press, 1928.

Jenner, Mrs. Henry. *Christian Symbolism.* London: Methuen, 1910.

Kemp, Martin. *Leonardo on Painting.* New Haven, CT: Yale University Press, 2001.

Kornbluth, Genevra. *Engraved Gems of the Carolingian Empire.* University Park, PA: Pennsylvania State University Press, 1995.

———. "Active Optics: Carolingian Rock Crystal on Medieval Reliquaries." *Different Visions: A Journal of New Perspectives on Medieval Art* 4 (2014): 1–36.

Levine, Robert. "Patronage and Erotic Rhetoric in the Sixth Century: The Case of Venantius Fortunatus." In *Words of Love and Love of Words in the Middle Ages and the Renaissance*, Medieval and Renaissance Texts and Studies, edited by Albrecht Classen, no. 347. Tempe, AZ: Arizona Center for Medieval and Renaissance Studies, 2008.

Lindberg, David. *Theories of Vision from Al-Kindi to Kepler.* Chicago: University of Chicago Press, 1976.

Macfarlane, Alan, and Gerry Martin. *Glass: A World History.* Chicago: University of Chicago Press, 2002.

Mack, Rosamond. *Bazaar to Piazza: Islamic Trade and Italian Art, 1300–1600.* Berkeley, CA: University of California Press, 2002.

Meiss, Millard. "Light as Form and Symbol in Some Fifteenth-Century Paintings." *Art Bulletin* 27, no. 3 (Sept. 1945): 175–81.

Ogle, M. B. "The Classical Origin and Tradition of Literary Conceits." *The American Journal of Philology* 34, no. 2 (1913): 125–52.

Pastan, Elizabeth Carson. "Glazing Medieval Buildings." In *A Companion to Medieval Art*, edited by Conrad Rudolph, 443–65. Oxford: Blackwell Publishing, 2006.

Pliny the Elder. *The Natural History*. Vol. 10. Translated by D. E. Eichholz. Cambridge, MA: Harvard University Press, 1962.

Roberts, Michael John. *The Humblest Sparrow: The Poetry of Venantius Fortunatus.* Ann Arbor, MI: University of Michigan Press, 2009.

Rogers, J. M. "European Inventories as a Source for the Distribution of Mamluk Enamelled Glass." In *Gilded and Enamelled Glass from the Middle East*, edited by Rachel Ward, 69–73. London: British Museum Press, 1998.

Romberg, Helen. "The Fatimid Treasury: Content and Function." M. Phil. thesis, University of Oxford, 1985.

Shalem, Avinoam. *Islam Christianized: Islamic Portable Objects in the Medieval Church Treasuries of the Latin West.* Rev. 2nd ed. Frankfurt am Main: Peter Lang, 1998.

———. "Some Speculations on the Original Cases Made to Contain Enamelled Glass Beakers for Export." In *Gilded and Enamelled Glass from the Middle East*, edited by Rachel Ward, 64–68. London: British Museum Press, 1998.

———. "Islamic Rock Crystal Vessels—Scent or Ampullae?" *Bamberger Symposium: Rezeption in der islamischen Kunst,* Beiruter Texte und Studien 61 (1999): 289–99.

Stern, E. Marianne. "Glass and Rock Crystal: A Multifaceted Relationship." *Journal of Roman Archaeology* 10 (1997): 192–206.

Stiaffini, Daniela. *Il vetro nel medioevo: Tecniche, strutture, manufatti.* Rome: Fratelli Palombi, 1999.

Tait, Hugh. "The Palmer Cup and Related Glasses Exported to Europe in the Middle Ages." In *Gilded and Enamelled Glass from the Middle East*, edited by Rachel Ward, 50–55. London: British Museum Press, 1998.

de Troyes, Chrétien. *Cligés*. Translated by Burton Raffel. New Haven, CT: Yale University Press, 1997.

Uboldi, Marina. "Diffusione delle lampade vitree in età tardoantica e altomedievale e spunti per una tipologia." *Archeologia Medievale* 22 (1995): 93–145.

Verità, Marco. "Analyses of Early Enamelled Venetian Glass: A Comparison with Islamic Glass." In *Gilded and Enamelled Glass from the Middle East*, edited by Rachel Ward, 129–34. London: British Museum Press, 1998.

de Voragine, Jacobus *The Golden Legend or Lives of the Saints*. Vol. 6. Edited by F. S. Ellis and Translated by William Caxton. London: J. M. Dent and Co., 1900.

Webb, Diana. *Medieval European Pilgrimage, c. 700–c. 1500*. Gordonsville, VA: Palgrave Macmillan, 2002.

Whitehouse, David. *Medieval Glass for Popes, Princes, and Peasants*. Corning, NY: Corning Museum of Glass, 2010.

Wilkinson, John. "Christian Pilgrims in Jerusalem during the Byzantine Period." *Palestine Exploration Quarterly* 108, no. 2 (1976): 75–101.

5

Transparent Glass in the West

Pietro Lorenzetti, Naddo Ceccarelli, and Others

As described in the preceding chapter, the Latin West had expanded its glass production over the course of the thirteenth century largely due to the influence of glassmakers and techniques from Islamic territories. As glass became more widely available, Italian devotional artworks increasingly incorporated various types of vitreous media. The primary case studies analyzed here—a group of trecento reliquaries that use transparent glass as small windows—suggest that artists may have consciously manipulated glass in order to resonate with the period's conception of vision and provide their viewers with highly effective devotional tools.

Glass was not just used for devotional art, however, and it is important to consider the reliquaries within the larger context of this medium's other functions. The rise of glass-making capabilities in late-thirteenth-century Europe led to an increasing number and variety of instruments that fostered new vantage points of, and ideas about, the natural world. For instance, glass lenses, in the form of either magnifying glasses or eyeglasses, extended one's ability to see as they aged, while glass windows protected one from the natural elements, but also allowed for a constant vista into the surrounding land and more light to flood into the interior.[1] Alan Macfarlane and Gerry Martin have argued that, when used in such ways, glass functioned as a "thinking tool," that is, glass, specifically transparent glass, provided new types of visual experiences and the contemplation of this new information allowed for advances in fields from medicine to mathematics to philosophy.[2]

The ability to access new types of visual information made possible by glass optical devices coincided with a shifting attitude towards the observable world. This period saw the burgeoning of what has been traditionally categorized as the Renaissance mentality—specifically, a growing interest in Aristotle's notion that sensory experiences are valid forms of information and a better understanding of the natural world can be achieved through empirical observation and experimentation. Characteristic of this time, such ideas were not isolated within the domain of the natural scientist or philosopher. Rather, they also permeated certain aspects of religious life and in fact, sometimes the churchmen and scientists were one and the same. Pope John XXI (r. 1276–1277), the so-called ophthalmologist-pope, wrote a textbook on caring for the eyes, *Liber de oculo*, which drew from ancient and Arabic knowledge and was consulted heavily at Salerno's medical school.[3] Perhaps the most famous instance of this outlook is the figure of Saint Francis of Assisi, who viewed the physical world as a reflection of God's divine grace and worthy of study and contemplation. This new approach to nature was a dramatic departure from that of earlier attitudes and the position of other groups like the Cathars, who understood the natural world as a potentially harmful imitation of and distraction from the heavenly ideal.[4] As a result, many individuals associated with the Order of the Friars Minor wrote about optical phenomenon in relation to their spirituality.

Artists also referenced optical theories in their art. A few well-known examples of artworks that reference optical ideas include Simone Martini's famous *Annunciation* from 1333, which seems to evoke the linear propagation of light with its rays of light emanating from the haloes in all directions as straight lines; Ambrogio Lorenzetti's roughly contemporary *Effects of Good Government in the City and Countryside,* which is thought to illustrate the theory that distant objects are slightly darkened because there is more space between them and the viewer; and Taddeo Gaddi's *Annunciation to the Shepherds* in the Baroncelli Chapel which depicts one of the early modern period's earliest representations of shadows and, as discussed in Chapter 2, may have been influenced by the artist's experience viewing a solar eclipse.[5]

Building on the premise that artists were investigating optical theories and the ideas discussed in the previous chapter (i.e., the eastern origins of optical treatises, glass reliquaries from the Holy Land, and the birthplace of glass in the eastern Mediterranean), what follows here illustrates the continued influence of Levantine glass and Arabic optical theory on the Latin West as well as several important connections between vitreous media and the sense of sight. Like the other types of glass discussed in earlier chapters, the current one outlines how the artistic applications of transparent glass were inextricably linked with religious symbolism, optical theory, and powerful visual encounters.

Reliquaries with Glass: The Case Studies

Examples of fourteenth-century Italian reliquaries that incorporate transparent glass are numerous and vary in function, format, and decoration. The common trait uniting the group under investigation here is the use of glass for, what I refer to as, "relic windows," that is, small panels of glass that act as display cases for relic fragments. While it was not a new development for reliquaries to feature visible relics, the collection of works discussed below use glass to mediate the relationship between relic and viewer in a relatively consistent manner which was not seen prior to the fourteenth century.

The group of reliquaries presented here is organized into three subgroups based on where the artists were working. The first subgroup consists of ten objects which were all made by Sienese artists. It includes the following examples: Naddo Ceccarelli's *Reliquary Tabernacle with Virgin and Child* (Figure 5.1); Lippo Vanni's *Reliquary with Virgin and Child with Saints* (Figure 5.2); Lippo Vanni's *Reliquary Triptych* in the Vatican Collection (inventory no. 40224); Lippo Vanni's *Reliquary Tabernacle* in the Società degli Esecutori delle Pie Disposizioni of Siena;[6] Francesco di Vannuccio's *Reliquary Tabernacle* in the Fondazione Monte dei Paschi of Siena (inventory no. FMPS101551–2642); Francesco di Vannuccio's *Reliquary with Madonna and Child and Saints John the Baptist, Paulinus of Nola and Louis of Toulouse* in the Museo Civico of Montepulciano; Bartolo di Fredi's *Reliquary Triptych with the Annunciation, Saint Ansanus, the Adoration of the Magi, and the Crucifixion* in a private collection in London from around 1370; and Pietro Lorenzetti's double-sided reliquary which is now divided, with the *Reliquary Tabernacle with Madonna and Child* in the Berenson Collection (Figure 5.3) and the *Reliquary Tabernacle with Enthroned Christ* now in a private collection in Rome.[7] Two other incomplete examples are found in the Cleveland Museum of Art (inventory no. 1978.26) and the Metropolitan Museum of Art (Figure 5.4), which, in their original condition, would have been very similar to other case studies in the group.

The second subgroup includes works by artists from Northern Italy and includes Simone dei Crocifissi's *Reliquary Triptych with Madonna and Child* in the Louvre (inventory no. D. L.1973–15); Simone dei Crocifissi's *New Testament and Apocryphal Scenes with Saints* in the Walters Art Museum (Figure 5.5), which

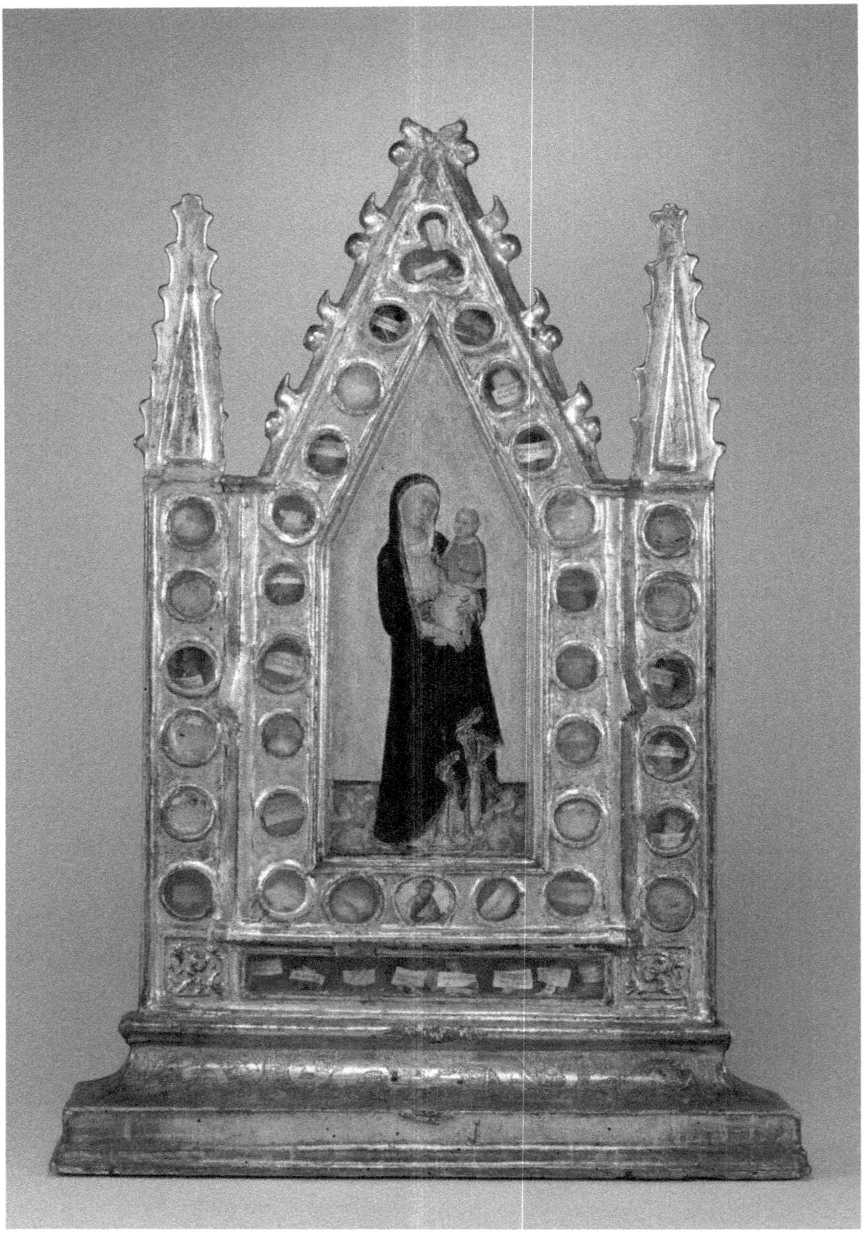

Figure 5.1: Naddo Ceccarelli, *Reliquary Tabernacle with Virgin and Child,* ca. 1350, The Walters Art Museum, Baltimore, Acquired by Henry Walters, 1920 (37.1159). Source: Public Domain, The Walters Art Museum (CC0).

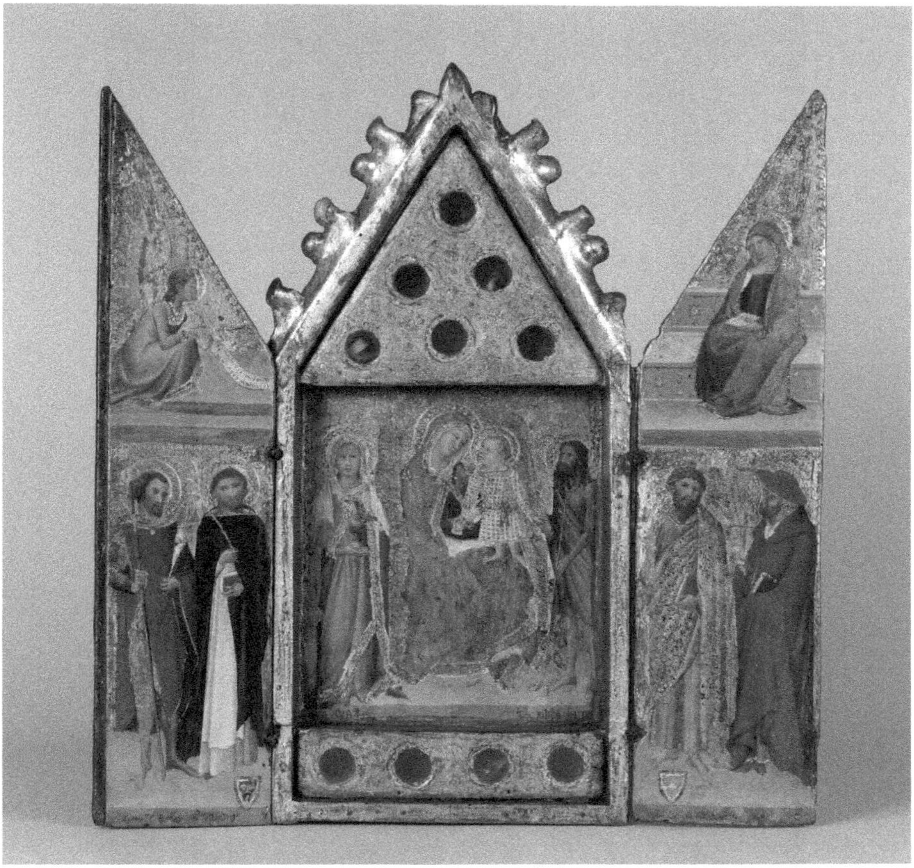

Figure 5.2: Lippo Vanni, *Reliquary Triptych with Virgin and Child with Saints,* ca. 1350–1359, The Walters Art Museum, Baltimore, Acquired by Henry Walters with the Massarenti Collection, 1902 (37.750). Source: Public Domain, The Walters Art Museum (CC0).

is currently missing a panel; Simone dei Crocifissi's *Crucifixion and Descent into Limbo* in a private collection; and Bartolomeo and Jacopino da Reggio's *Triptych with Crucifixion, Annunciation, and Saints* in the Pinacoteca di Brera in Milan (inventory no. 6019).

The third subgroup consists of reliquaries made in and around Umbria. It includes the *Reliquary Diptych with Virgin and Child with Saints* from the Victoria and Albert Museum by the Master of Sant'Alo' (inventory no. 19–1869) and several reliquaries discussed in the next chapter. The vast majority of this group is

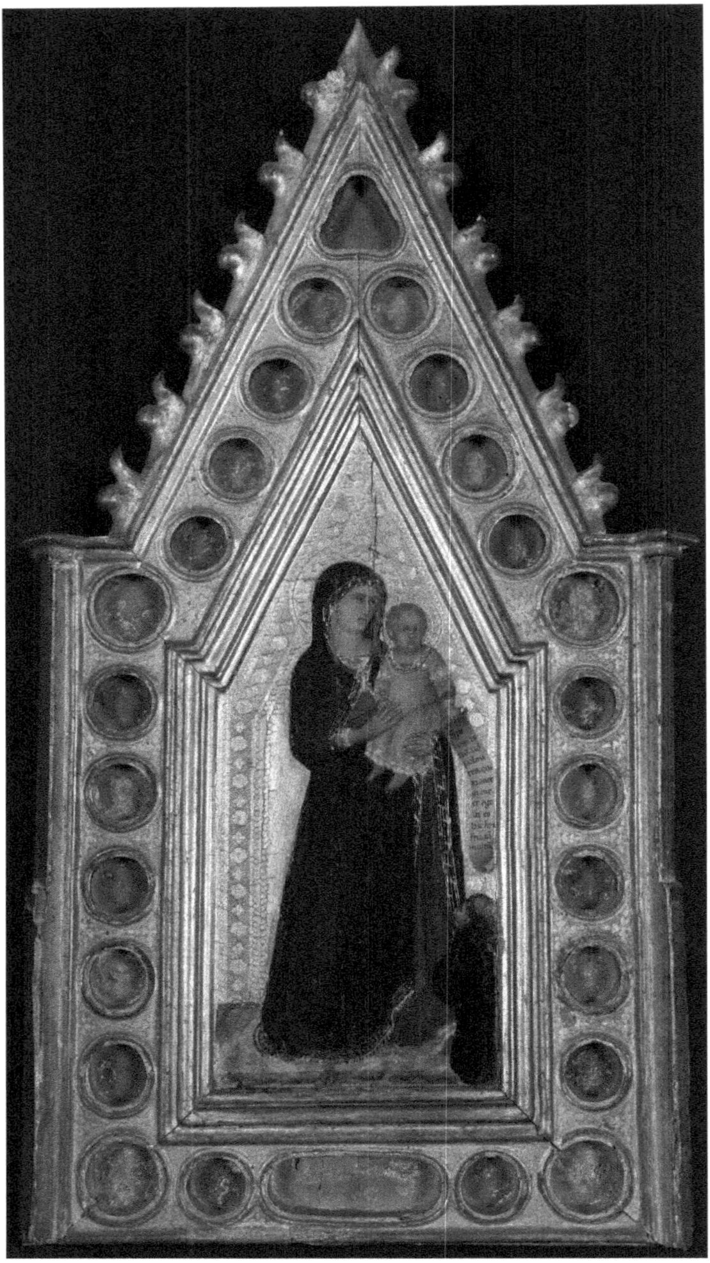

Figure 5.3: Pietro Lorenzetti, *Reliquary Tabernacle with Madonna and Child,* early 1340s, Florence, Villa I Tatti, Collezione Berenson, reproduced by permission of the President and Fellows of Harvard College. Photo: Paolo De Rocco, Centrica srl, Firenze.

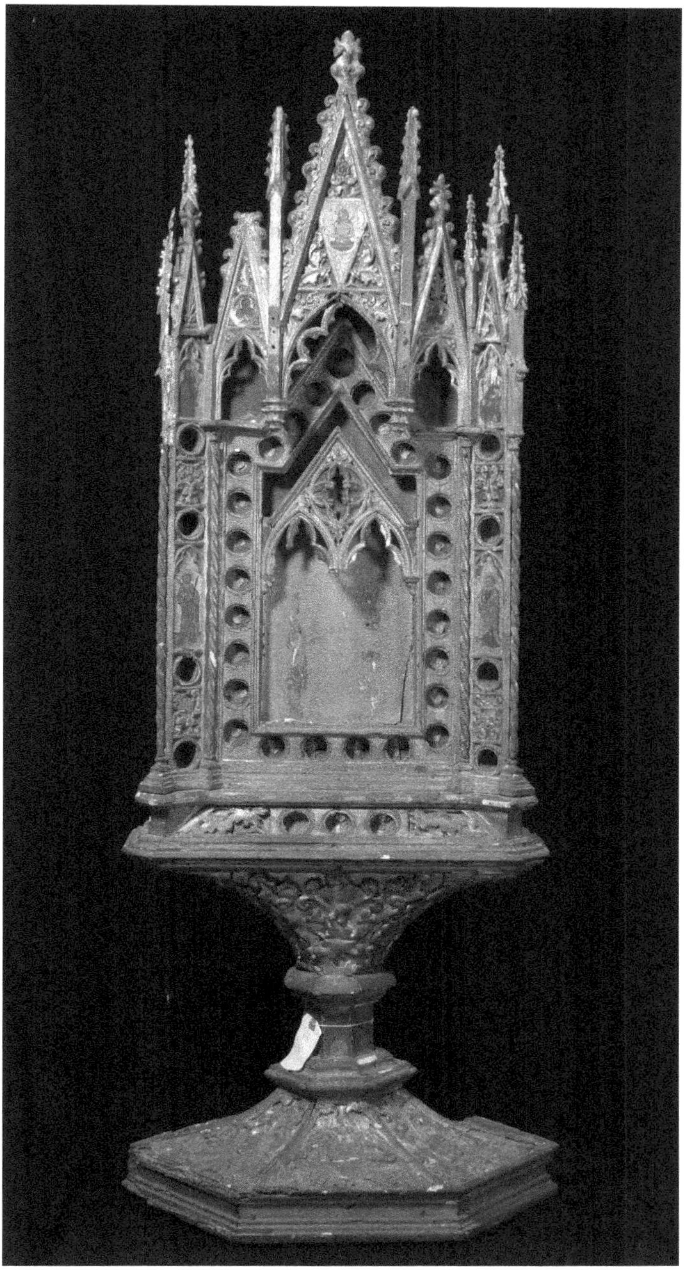

Figure 5.4: Sienese Workshop, *Reliquary Tabernacle*, 14th century, The Metropolitan Museum of Art, New York, Rogers Fund, 1918 (18.70.17). Source: Public Domain, The Metropolitan Museum of Art (CC0).

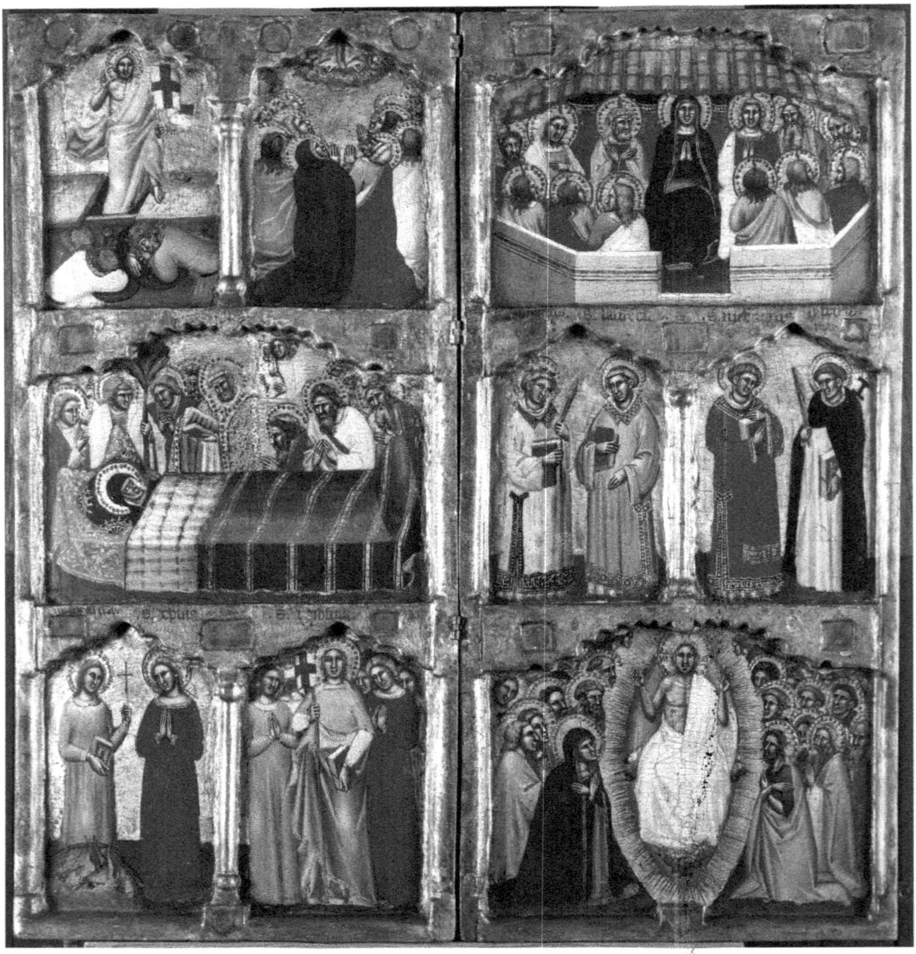

Figure 5.5: Simone di Filippo (Simone dei Crocifissi), *New Testament and Apocryphal Scenes with Saints,* ca. 1360–1370, The Walters Art Museum, Baltimore, Acquired by Henry Walters with the Massarenti Collection, 1902 (37.723). Source: Public Domain, The Walters Art Museum (CC0).

discussed in the following chapter because these reliquaries combine portions of transparent glass with *verre églomisé* glass, a style that produces such a dramatically different visual effect that it warrants a separate discussion.[8]

Aside from the unifying element of the relic windows, the group of case studies features a variety of formats. Some of the reliquaries are portable devotional devices in the form of diptychs, triptychs, or polyptychs, while others are free-standing tabernacles meant for public display or even processions. Some works feature painted imagery, while others include verre églomisé imagery. The subject matter is also eclectic. Scenes of the Virgin and Child and the Crucifixion are popular but one can also find less common themes such as Christ's descent into limbo and apocryphal stories. Even the format of the relic window is not replicated consistently throughout the group. Some reliquaries have—or once had—prominent, circular relic windows while others feature quatrefoil decoration framing the glass panel. In fact, the *oeuvres* of Lippo Vanni and Simone dei Crocifissi demonstrate that a single artist could in fact work with different types of relic windows, suggesting either changing fashions or a situation in which the artist did not dictate the design of the frame or relic window.

Reliquaries by Naddo Ceccarelli and Bartolo di Fredi in Context

Of all the trecento reliquaries listed, that by Naddo Ceccarelli now in the Walters Art Museum (Figure 5.1), has received the most recent scholarly attention because of its excellent state of preservation and the scholarly consensus regarding its attribution.[9] Like several other reliquaries discussed here, it features naturalistic paintings of holy figures surrounded by small, round, relic windows.[10] Ceccarelli's panel is characteristic of his work in its extensive use of gilding and the architectural format of the frame, which reflected the style of contemporary goldsmiths such as Ugolino di Vieri's *Reliquary Head of Saint Savinus,* and the influence of Simone Martini's delicate yet naturalistic style.[11] The Virgin Mary's refined elegance is emphasized by her elongated figure, the linear treatment of her flowing drapery, and the overall vertical nature of her standing pose, which is reinforced by the panel's pointed gable and pinnacles.

The Madonna gives the impression of a divine queen, however, at the same time, she seems grounded in the physical world. The gilded background, lacking any earthly referent, was firmly outside the realm of human experience and therefore evocative of a celestial, supernatural space. Mary seems to float or hover in this ethereal setting because of the way her dark blue garment is almost entirely silhouetted against the golden void. Her feet, however, reveal a different reality. She stands in a slight s-curve and, as her voluminous drapery falls to the marbled floor, it pools into small piles. Such passages convey reasonable anatomy, the suggestion of gravitational forces acting on the figure and their garments, and the modeling of form with highlights and shadows, which are the hallmarks of a painter with naturalistic interests. The result of all this is an underlying tension between the heavenly elements and the more earthly ones.

C. Griffith Mann finds a similar interplay between things spiritual and material in the relationship between the glass roundels and the painted scene at the center, a dynamic he describes as "a confrontation between matter and spirit."[12] Indeed, reliquaries and relics were touch points where the material and spiritual worlds intersected. The relics were tangible evidence of a saint's past life as a human being and, simultaneously, symbolic of their transcendence of the human world into the divine one. Thus, fittingly for a reliquary, the juxtaposition inherent in the relics and central to Mary's saintly status is visually manifested in the formal elements of the panel—in the coexistence of a naturalistic style and an abstract one and in the collaborative effort of the many different materials, including paint, wood, gold, and glass. Such a medley of media could have inspired contemplation of the miraculous transformations made by Christ and the saints as they transitioned between spiritual and material realities.[13]

Not surprisingly, the church as an institution aided the faithful viewer as they navigated the complicated theology surrounding relics, something that is reflected in the very format of the reliquaries. There was an increasing availability of relics in the thirteenth and fourteenth centuries because of growth in pilgrimages and the relic trade.[14] In particular, one saw an influx of relics from Byzantium, where regulations over handling and dividing relics were more relaxed than in the West.[15] These trends caused concern for the European churchmen about potential forgeries and the way in which the relics were venerated.[16] In order to preserve the sacredness of the relics and ensure that they were not treated as magical talismans or protective amulets, the church instituted regulations. At the Fourth Lateran Council of 1215, for instance, the church declared that relics were not to be worshipped unless they were in an appropriate container and sanctioned by the authority of a bishop.[17]

Reliquary tabernacles, like the one by Naddo Ceccarelli, offered an elegant solution to the church's growing concerns regarding relic veneration. They ensured that the display of the relics met the mandate that relics needed to be contained while maintaining the compelling aspects of the visual connection. This arrangement also allowed the church to more closely control how, when, and where one interacted with the relics. Set within a gilded wooden panel and elaborate framework, these relics could not be worn around one's neck and treated as apotropaic devices or protective amulets.[18] Furthermore, like many of the other trecento reliquaries featuring relic windows, Ceccarelli's panel is shaped like an aedicule, a clear reference to the design of many churches at this time.[19] Eliot W. Rowlands argues that the compositional similarity between the reliquary and the shape of a church was intentional because the reliquary was meant to symbolize the church with the Madonna serving as the gate to heaven, the *porta caeli,* and the relics representing the family of saints within the church.

The reliquaries' reference to church architecture underscores the ecclesiastical validation of the relics and saints. Positioned at the periphery of the reliquary, the relics function like symbolic walls of the church. In a manner analogous to the architectural ribs of the Gothic cathedrals, these fragments of bones and stones support the reliquary's pointed arches and pinnacles above the Madonna and Child.[20] The display of these reliquary panels on an altar table within an actual chapel would have further ensconced the saints and their relics within the authority of the church as a physical structure and, in turn, as an institution. The church's official approval of these relics was certified by the presentation of the relics' authentics, or identifying labels, and other layers of legitimacy such as hagiographies, ritual processions or ceremonies, and feast days.

Like Ceccarelli's tabernacle, Bartolo di Fredi's *Reliquary Triptych with the Annunciation, Saint Ansanus, the Adoration of the Magi, and the Crucifixion* also visually solidifies the relationship between the relics and church.[21] The glass roundels symbolically function as structural support for the frame and surround the central scenes, almost infusing it with a sense of physicality. Because this object is a triptych, the side panels could be opened or closed like the doors of a church. When opened, the pinnacles of the side panels flanked the larger, central pointed gable in a manner very similar to fourteenth-century church façades like the cathedrals of Siena and Orvieto. Additionally, one finds these same architectural elements—three pointed arches—repeated again within the ivory relief at the center of the reliquary. These multiple layers of reference to church architecture appear to advertise the relics' many forms of institutional review, control, and support.[22]

When Bartolo's reliquary was opened, the paintings of the Virgin Annunciate and the Angel Gabriel on the wings would have framed the central panel's scenes of the Adoration and Crucifixion depicted in the ivory panels, and this would have further emphasized the connection between the reliquary and the church. The conversation between the angel and Mary was the inception of Christ's life on earth and this event flanks the central narrative scenes in the ivory panels. The left panel of ivory features a scene from Christ's early years, the Adoration of the Magi, while the ivory panel on the right shows the Crucifixion, his departure from the physical world. Positioned like this, Christ's entire life, from his early childhood to his death, was encompassed within the interaction between Gabriel and the Virgin Annunciate in both iconographic and formal terms.

Such a phenomenon was not unique to this reliquary; the use of the Annunciation as a visual—and symbolic—entryway into scenes depicting the New Testament was also found in contemporary church architecture. The most famous early-fourteenth-century example is Giotto's frescoed version in his Arena Chapel, where the Angel Gabriel and the Virgin Mary flank the altar and initiate the registers of narrative scenes from Christ's life on the adjacent walls. A similar situation, executed in sculpture, also exists at the Baroncelli Chapel in Santa Croce, Florence.[23] Flanking the entry portal to the Baroncelli chapel are Giovanni di Balduccio's sculptures of the Angel Gabriel, at the left, and the Virgin Annunciate, at the right, set high above viewers on pedestals protruding from the chancel walls.[24] These sculptures date to around 1327 and Giotto's example dates to about 1305, suggesting that there was an established tradition by the time Bartolo was making his reliquary. Further indicating Bartolo's interest in this motif is the fact that he experimented with spatial issues within depictions of the Annunciation in his later commissions.[25]

Bartolo's reliquary, like many of the others with relic windows, provided the viewer with a plethora of meditative avenues because of its complex iconography, the combination of various types of media, and the way the parts relate to one another. Gazes and gestures guide the viewer from God the Father down to the Angel Gabriel and then over to the Virgin. Then, when coming to the center panel, the viewer works their way up from the stories of Christ's life to Saint Ansanus, whose prominent gaze upward points to the relics and, ultimately, heaven. In following these lines and what they represent, this reliquary could have served as a portal for the spiritual imagination to contemplate many different aspects of the faith. By using one's earthly sight to attain spiritual insights, the viewer entered an imaginative journey from earthly realities to spiritual beliefs.

Relic Windows: Precedents and Influences

To better understand the relic windows in the aforementioned reliquaries, it is necessary to investigate some of the complex factors that contributed to the West's growing interest in putting their relics on display.[26] As Chapter 4 describes, at the same time as glass production was escalating, early Christian pilgrims making the journey to the *loca sancta* brought back holy souvenirs made from glass, or recorded this practice in their pilgrimage accounts. Another potential source of inspiration may have been Fatimid crystal pilgrim flasks, which were typically integrated into a larger reliquary ensemble as discussed in the last chapter. A third important influence came from the east in the form of Byzantine reliquaries.[27] Although Byzantine reliquaries do not generally feature an extensive use of glass, they did offer visual access to the relics and thus may have contributed to the desire to see the relics. *The Limburg True Cross Reliquary (Staurotheke)* in the Diözesanmuseum, Limburg for instance, allowed the privileged viewer to slide open the lid to reveal the sacred wood of the true cross and then open the smaller doors for access to numerous secondary relics. Another method of display in the Byzantine tradition was to leave the relic largely unadorned, as seen in the *Skull Relic of Saint James the Younger* from before 1204 and now in the Domschatzverwaltung, Halberstadt. When removed from its case, the viewer had direct visual access to the skull, which has been embellished and decorated but left mostly visible. Both of these reliquaries entered Europe after the Sack of Constantinople in 1204, making them particularly informative examples for an analysis of the Byzantine influence on western reliquaries from the trecento.

The relationship between the Byzantine Empire and the Latin West is a complex one that reaches as far back as antiquity, when their cultural heritages were united within the Roman Empire. In the centuries following the fall of Rome, the Byzantine Empire flourished while the western territories experienced a period of political instability beginning in the fifth century and lasting until Charlemagne. Theological differences came to fruition in 1054 with the East–West schism but relations were maintained for various reasons such as a mutual interest in the Holy Land, the shared value of Christian relics, a desire to keep the Biblical holy sites in the possession of Christian forces, and a common enemy in Islam.

As the Byzantine emperor possessed many of the most important relics of the Passion of Christ, he was the gatekeeper to the Latin West's access prior to the 1204 Sack of Constantinople. The emperor capitalized on this powerful position by sending western dignitaries relics as diplomatic gifts, determining whether a visitor to Constantinople was granted access to them, and presenting European hosts with relics when he traveled abroad.[28] However, the situation changed dramatically after 1204 when what began as an Italian-led attempt to wrest the holy city of Jerusalem from the Muslims ended with the looting of the Imperial Treasury of the Eastern Christians. The wide-ranging benefits gained from the 1204 Sack of Constantinople included booty such as the famous bronze horses that were to grace the top of Saint Mark's, an impressive amount of tesserae, and numerous glass and crystal luxury objects.[29] Many important reliquaries were part of this loot and, therefore, were removed from Byzantium and brought to the Latin West. Because of their revered status, when these objects entered western collections, they would have been viewed as highly valued and thus very influential.

A particularly informative moment of artistic exchange is revealed when the aforementioned *Limburg True Cross Reliquary*, which entered the west as booty from the Sack of 1204, is compared with the *Cross Reliquary at Trier* from the first half of the thirteenth century.[30] It is immediately evident that the reliquary at Trier closely emulated the overall composition of the reliquary at Limburg, borrowing the scale and central placement of the cross as well as the checkerboard setting of various small panels containing many more relic fragments. When comparing the two, however, it is important to note their differences. In the Byzantine example, the smaller relics are accessed by opening the doors while the reliquary at Trier displays the relics behind pieces of transparent crystal. In the Byzantine work, the relics are either enclosed behind the small golden doors or exposed, while in the work from Trier, the relics are simultaneously contained and revealed in accordance with church regulations.

The Cross Reliquary at Trier demonstrates how a European artist might have fused Byzantine influence with western artistic traditions and religious customs. The use of the transparent glass for relic windows was not yet widespread in the thirteenth century but the desire to view relics through a translucent or transparent material such as rock crystal was already well established. For example, *The End of a Reliquary Shrine with the Triumphant Christ* in the Walters Art Museum (inventory no. 57.519) from the Mosan region[31] features thin sheets of translucent horn in a similar manner, while *The Reliquary of Saint Francis* in the Louvre (inventory no. OA 4083) uses rock crystal.

Another association that may have inspired the increasing use of glass in reliquaries was this medium's ancient associations with venerated remains. There were many glass vessels to be found throughout the Roman catacombs, which is not surprising considering that Ancient Rome had one of the most prolific glass industries. The vitreous medium was used for everything from dishware to windows.[32] Glass was also used to make large storage vessels, some of which were used as funerary urns. These jars came in a variety of formats, some with handles, some without handles.[33] They are made of thick blue-green glass, which is the naturally occurring color of this glass; however, some colored glass jars have been found. Oftentimes they have lids that served as funnels through which one could pour an offering of liquid. A second type of ancient Roman glass associated with burial sites is the sandwich gold glass, called this because gold leaf designs are sandwiched between two sheets of transparent glass. As discussed in Chapter 3, gold glass roundels featuring pagan, Jewish, and Christian motifs typically decorated the base of a drinking vessel or bowl and after their owner's death, these functional objects were strategically broken to free the gold glass roundels and impressed into the cement wall of the deceased's tomb in the catacombs (Figure 3.3).[34]

In addition to their association with the holy sites of Rome, the ancient gold glass roundels may have also carried the aura of a secondary relic, that is, an object sanctified by physical contact with the body of a holy figure. Not only did the glass roundels press against the wall containing the remains of the deceased martyrs but, also, while he or she was still living, the glass object would have come into physical contact with the saint when the bowl or plate used. Considering this, it is not difficult to imagine that an encounter with these gold-glass roundels could have held spiritual importance for the medieval Christian and may have influenced their reception of verre églomisé in Christian reliquaries.

Both the gold glass roundels and the glass funerary urns were likely seen by many medieval and Renaissance pilgrims. The Roman catacombs were popular sites for Christians searching for sacred tombs such as that of Saint Peter, especially when pilgrimages to Rome increased during the fourteenth century. Seen as both a response to already occurring pilgrimages and an impetus for more in the future, Pope Boniface's papal bull, the *Antiquorum habet fida relatio*, codified a generous indulgence policy for the Roman Jubilee in 1300 that awarded pilgrims to Rome a spiritual reward similar to that offered to crusaders.[35] Inspired by the promise of salvation, early modern Christians traveled to Rome to pray at, among other sacred sites, the burial grounds of the venerated Early Christian martyrs. Thus, by the time the trecento reliquaries with relic windows were being made, there may have already been strong connections between Christian relics and the vitreous medium.

Relic Windows in "An Age of Vision"

The desire to make visual contact with relics was just one facet of a larger cultural phenomenon. Over the course of the thirteenth and fourteenth centuries, Christians sought out various ways to employ their sense of sight in pursuit of spiritual understanding and enlightenment. The result was a culture that was interested in optical theory and putting the study of this theory into practice. Many scholars have commented on the increasingly visual nature of late medieval society; one of the earliest was Johan Huizinga, who understood the trecento's interest in visual information as a decline in culture of sorts.[36] Writing about fifty years after Huizinga, Alfred Crosby takes a different approach and correlates the growing prominence of the visual sense with the need to quantify goods and other entities in his book, *The Measure of Reality: Quantification and Western Society, 1250–1600*.[37] Crosby outlines a shift from a culture dominated by sounds—where holy scripture is delivered by sermon, decrees are announced at public gatherings, and time is kept according to the ringing of the town bell—to one where the sense of sight provides the most important types of information—where educated Christians read devotional treatises or prayer books, official documents are written and signed, and time is told by a clock or hourglass.

Hayden Maginnis's art historical analysis, identifies and traces an emphasis on visual experiences throughout the fourteenth century's seminal works of art, a period which he calls "an age of vision." Maginnis demonstrates how a penchant for visual media manifested itself in various guises, from literature to painting.[38] For instance, it was a visual encounter with a painting that was the primary catalyst for such important events as Saint Francis's divine call to action and Boccaccio's poem, the *Amorosa visione*. Poems by Dante and Guido Cavalcanti also make numerous references to detailed visual descriptions, almost imploring their readers to envision their fictive worlds within the mind's eye.[39] When considering the popularity of visible relics and glass relic windows, it is important to cast a wide net and examine the ways these objects resonated with their larger, visually inclined cultural context. Doing so reveals that an interest in engaging with the world by means of vision was not unique to relic worship, but rather many aspects of trecento culture and religion were geared towards one's sense of sight.

Mirroring trends in relic veneration, the sacrament of the Eucharist also began to cater to a more visually inclined audience. During the 1230s there was intense enthusiasm for the Elevation of the Eucharist, the moment during the mass when the priest raised the transubstantiated host to the crowd, showing them the miraculous moment of transition from piece of bread to the body of Christ. The act of viewing the Elevation had become such a priority for the congregation that many faithful Christians ran to several different churches in a single day in order to see as many rituals as possible.[40] It reached the point that some church leaders worried the public was no longer interested in consuming the sacrament but content to solely view it.[41] During the fourteenth century, the frantic need to rush from Elevation to Elevation subsided but the desire to see the Eucharist did not, as demonstrated by the proliferation of trecento monstrances that displayed the host behind rock crystal in an analogous manner as the relic windows.[42]

Both churchmen and scientists associated with religious orders found vision beneficial to cultivating divine understanding. Alexander of Hales (ca. 1170–1245), a Franciscan theologian writing at the same time as the Elevation frenzy, advocated interacting with the Eucharist visually. Following ideas found in Aristotle's recently translated works, Alexander believed that a search for knowledge originated with one's senses.[43] He reasoned that Eucharist's benefits were of a spiritual and immaterial nature and, therefore, the sense of sight was the most appropriate sense for encountering the host, as it was the least material of the five senses.[44] Roger Bacon (ca. 1120–1292), one of Alexander's contemporary fellow Franciscans, also held the sense of sight and the ideas of Aristotle in high esteem. Bacon described optics as "the flower of the whole of philosophy" because it sheds light on all the other scientific pursuits. For Bacon, one can "understand nothing fully unless its form is presented before our eyes,"[45] and this applied to the pursuit of both heavenly and earthly knowledge.[46]

It is not a coincidence that both Alexander of Hales and Roger Bacon drew from Aristotle, or The Philosopher, as he was called at this time. Aristotle's ideas began to permeate European thought around the middle of the thirteenth century when the majority of Ancient Greek and Arabic literature was translated to Latin.[47] The arrival of Aristotelian ideas about the nature of reality and the sense of sight altered the trajectory of visual theory from what had been an essentially Neoplatonic system of extramission vision—where vision results from rays extending outward from the eye—to Aristotle's version of intromission—where the eye receives information from the perceived object.

The theory of extramission vision, along with other Platonic ideas about reality, had dominated the Latin West due to the influence of Augustine (354–430) from the fourth century through the end of the thirteenth century when Aristotelian ideas began to supplant them.[48] William of Conches (ca. 1100–54) and Adelard of Bath (fl. 1110–40) were some of the first to deviate from pure extramission vision when they described in detail how visual rays not only leave the eye, but, upon making contact with the object, they returned to the beholder.[49] Robert Grosseteste (ca. 1175–1253) continued this trend in the following century. His theories are still basically Platonic but he deviates by declaring his debt to Arabic sources.[50] Albertus Magnus (ca. 1200–80) and, to a greater extent, Roger Bacon (ca. 1220–92), mark a significant turning point in western optics, one based more solidly on the intromission theories found in Aristotle and Arabic scientists such as Alhazen.[51]

Central to Aristotle's theory, and important for the present discussion about glass, is the way in which the information is transferred from the object to the eye in the intromission model. Aristotle argued that vision occurs when the perceived object emits images of itself, known as *eidola*, into surrounding transparent media. As these entities make their way towards the eye, they alter the transparent medium until they reach the eye's surface.[52] Such a visual theory is less concerned with the direction of the visual rays and more focused on the integral role played by the transparent medium between the eye and object.[53] As Aristotle described,

> The evidence for this is clear; for if one puts that which has colour right up to the eye, it will not be visible. Colour moves the transparent medium, *e.g.*, the air, and this, being continuous, acts upon the sense organ. Democritus is mistaken in thinking that if the intervening space were empty, even an ant in the sky would be clearly visible; for this is impossible. For vision occurs when the sensitive faculty is acted upon; as it cannot be acted upon by the actual colour which is seen, there only remains the medium to act on it, so that some medium must exist; in fact, if the intervening space were void, not merely would accurate vision be impossible, but nothing would be seen at all.[54]

Thus, for Aristotle's and his adherents, a great deal of importance was now placed on the medium between the eye and the observed object. And, importantly, for visual contact to be successful, this medium needed to be transparent.

Though it seems a logical candidate, I am not aware that either Aristotle or Bacon specifically mentioned colorless glass in reference to this transparent medium. Dante, however, does seem to have made such a connection. In *Convivio* III, ix, he describes a visual process very similar to that of Aristotle using glass as an example of the transparent medium, noting, "These visible things … enter the eye—I do not mean the things themselves, but their forms—through the diaphanous medium, not actually but mentally, as through transparent glass."[55] The fact that Dante was familiar with current trends in visual theory and described them in vernacular prose suggests that this type of optical theory should not be seen as entirely inaccessible to the general educated public.[56]

Viewing reliquaries such as that by Naddo Ceccarelli (Figure 5.1) with an Aristotelian model of vision in mind, as Dante and like-minded viewers might have done, a fourteenth-century observer could have interpreted the relic window as a stable, physical manifestation of the visual process, an affirmation that the relic was successfully sending forth its *eidola* into the surrounding medium, through the glass, and ultimately to the viewer's eye. The fact that the eye's anatomy was thought to be made of crystalline and vitreous elements, as discussed in the last chapter, could have inspired even more confidence that the visual impression of the relic was received and processed by the viewer's eye and mind. Though it was not the seat of vision, the glass-like humor was a key component in facilitating the visual process because it was the intermediary between the observed object and the viewer's body. In a similar way, the prominent round, glass relic windows in the reliquary facilitated one's connection with the relics, which in turn are representations of the physical bodies of these saintly intercessors. Thus, the function of glass as a mediator, or a transmitter, pervaded an encounter with the reliquaries.

Pietro Lorenzetti's *Reliquary Tabernacle* in Context

Such context is particularly important when evaluating reliquaries such as Pietro Lorenzetti's aforementioned *Reliquary Tabernacle with Madonna and Child* (Figure 5.3) and *Reliquary Tabernacle Christ Enthroned,* the two panels of which functioned as the front and back of a single object.[57] Pietro's reliquary is one of the most important within its group because it is the earliest of the Sienese tabernacles, dating to the early 1340s, and therefore may have served as the prototype for the others. Such a hypothesis is logical given Pietro's prominent reputation in early fourteenth-century Siena. By mid-century, Pietro, along with his brother, Ambrogio, was one of the most prominent artists in Siena and the surrounding area. Pietro was commissioned to make many large-scale altarpieces for such institutions as Pieve di Santa Maria in Arezzo, San Marco in Cortona, San Niccolò al Carmine in Siena, and, most importantly, the famous altarpiece depicting the birth of the Virgin for the Saint Savinus chapel in the Cathedral of Siena. Also supporting the notion that Pietro's reliquary may have influenced the others is the fact that common traits found in many examples, namely the vertical arrangement of the composition, the pointed gable, the relic windows' placement along the periphery, and the naturalistically painted figures with gilded background, are all present in this artwork.

Furthermore, it is likely that among all the artists mentioned in this chapter, Pietro Lorenzetti was especially sensitive to the symbolic potential of relic windows, a factor that may further support his role as an innovator of the relic window format. Pietro Lorenzetti's participation within the network of glass and other, related optical issues can be found throughout his oeuvre. About a decade before working on his reliquary, roughly around 1328, Pietro painted two transparent glass vessels in the scene of *Hermits at the Fountain of Elijah* for predella of the now dismantled altarpiece for San Niccolò al Carmine in Siena.

Additionally Pietro's frescoes in the Lower Church of San Francesco in Assisi are often cited for their revolutionary naturalism and illusionistic play between real space and pictorial space, making them good examples of his growing interest in art that reflects empirical observation of the natural world. But, these frescoes also suggest that Pietro may have understood the ability of glass, as a material, to resonate with important aspects of religious symbolism. The clue to this aspect of his style is revealed in the left transept of the Lower Church, where Pietro painted an illusionistic three-dimensional niche containing the items needed for celebrating the mass. Tucked away in this fictive cupboard, Pietro painted two transparent glass vessels, one of which holds the wine that was to be transformed into the blood of Christ during the ritual of the Eucharist (Figure 5.6).[58]

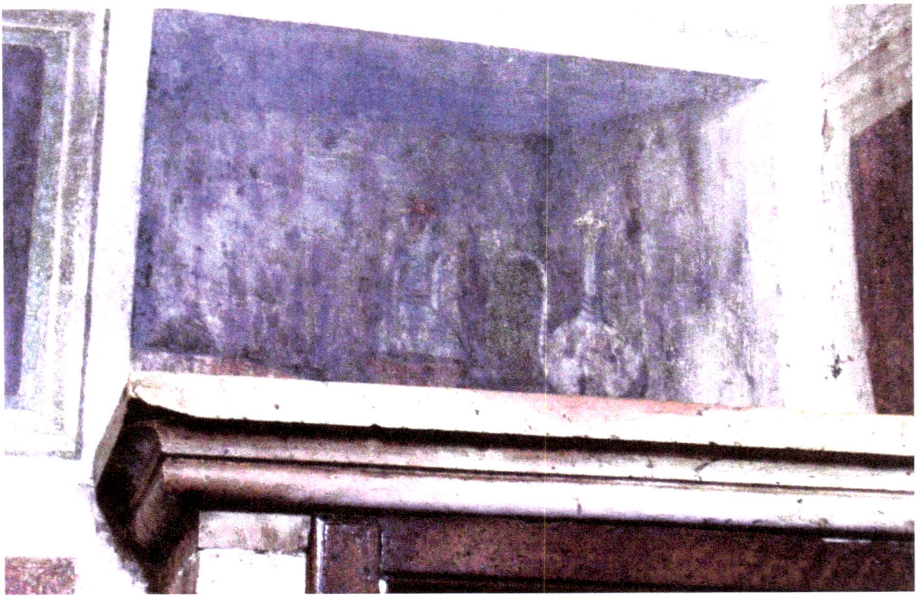

Figure 5.6: Pietro Lorenzetti, *Illusionistic Niche with Transparent Glass Liturgical Vessels*, ca. 1320, Left Transept, Lower Church, San Francesco, Assisi. Source: Author.

Glass's ability to both contain and reveal something as symbolically significant as the holy wine seems especially appropriate. As mentioned in Chapter 4, the association between the Incarnation of Christ and light traveling through glass without breaking had been well established by the time Pietro was working. It is not hard to imagine that, as he painted his fictive niche and depicted vessels related to the Last Supper and—thus ultimately, Christ's Passion and Resurrection—the artist contemplated the role played by glass in the analogy of the divine Incarnation. In other words, Pietro may have depicted the illusionistic transparent glass vessel as a type of visual meditation on transubstantiation. The unique qualities of glass—in both the depicted wine vessel and in the reliquary's relic windows—allowed the viewer to contemplate the paradoxical nature of Christ's miraculous passages. The belief that Christ was conceived by a virgin and, after his human death, rose to heaven from his sealed tomb were not only complex ideas but, correspondingly, complex topics to depict in a narrative art form. Treating such subjects in a more abstract manner was certainly a plausible approach. The possibility the unique properties of glass may have helped the artist, the viewer, or both navigate these contemplative maneuvers should not dismissed. It is not a coincidence, I argue, that Pietro's depiction of transparent glass liturgical vessels allowed one's gaze to penetrate through to the holy wine much in the same way one's eyes penetrated the relic windows in his reliquary, which was made only about 20 year after the frescoes.[59] Pietro Lorenzetti's other work in the Lower Church might suggest yet an additional way in which this prominent artist may have been engaged with the network of glass and, more specifically, with optical devices featuring roundels of glass which may have resonated with his relic windows in the reliquary.

Relic Windows, Mirrors, and Eyeglasses

In the vault of the Lower Church at Assisi, mere steps from Pietro's illusionistic niche with the depiction of transparent glass vessels, is the scene of the *Allegory of Obedience,* which is attributed to Giotto's workshop and dated to the same time as Pietro's frescoes at this location. In the vault a Franciscan friar kneels before a figure that appears to be Prudence, who holds a small, circular mirror out towards the friar. Thus, another important reference to optical devices, which were, at least some of the time, made using round panels of glass.[60] Two other mirrors from the first half of the trecento with which Pietro made have been aware are attributed to his brother Ambrogio. The first is featured in the fresco cycle depicting the *Allegory of Bad Government* in Siena's Palazzo Pubblico in the hands of a personification of Vanity, who gazes into it as she hovers above the figure representing tyranny. The second is held by the figure of Faith in the altarpiece depicting the *Virgin and Child Enthroned* from Massa Marittima.

Other fourteenth-century artists were also depicting mirrors in their artworks. Giotto painted one in his illustration of Prudence, as did Andrea da Firenze painted on in the throne of the *Triumph of Saint Thomas Aquinas* in the Spanish Chapel of Santa Maria Novella. A more in-depth analysis of these mirrors is found in the next chapter but they should be mentioned here too in order to establish that Pietro Lorenzetti and other trecento artists were familiar with mirrors and their optical properties and may have contemplated their implications for one's sense of sight.

References to visual theory and optical devices would have been well received by the Franciscan patrons of Pietro's reliquary panels and the frescoes in the Lower Church, who are prominently featured in the panels by way of the small figures kneeling in adulation and wearing the Franciscan habit. Saint Francis was at the vanguard of a shifting attitude towards the natural world because he placed spiritual value on the physical reality around him. This attitude is poetically expressed in his *Canticle of the Creatures*, a famous hymn originally written in the local Umbrian dialect that celebrates various aspects of the natural world such as the sun, moon, stars, and animals.[61] Francis's high regard for nature permeated his order and opened it up to the study of natural science and, in particular, optics.[62]

Not surprisingly, the Franciscan milieu at Oxford and Paris significantly contributed to one of the most influential early modern optical theorists: Roger Bacon, whose work was cultivated. As Timothy Johnson describes the situation

> To care for creation and embark on the journey into God requires close attention to the natural world. … [Bacon gave] voice to the Franciscan fascination with the myriad reflections of divine agency and purpose in the materiality of the earth.[63]

But Bacon was just one of many. Several other of the leading optical theorists were also affiliated with the Franciscan order.[64] John Pecham (ca. 1225–92), John Duns Scotus (ca. 1266–1308), Bartholomeus Anglicus (b. before 1203–72), Robert Grosseteste (ca. 1170–1253), and William of Ockham (ca. 1287–1347) all belonged to or maintained associations with the Order of the Friars Minor.[65]

As already mentioned in the discussion of the Lower Church frescoes, the Franciscan penchant for optics is also reflected in some of the order's most famous commissions. Many works of both literature and visual art made for the Order of the Friars Minor invoke powerful inner visions or give visual form to ideas that are unobservable. *The Meditations on the Life of Christ,* written around 1300 and attributed to Bonaventure, is a text devised to aid the devotional practices of a nun of the Poor Clares.[66] It vividly describes biblical events and implicates the reader as a witness by crafting a rich sensory experience with the goal of inspiring the reader to envision the events in one's mind's eye. The artistic programs at San Francesco in Assisi, along with the order's many other commissions, feature some of the period's most famous examples of naturalistic painting and complex narratives, many of which illustrate important moments in the life of the order's founder in a way that allows the viewer to imagine they are one with the painted world. The Franciscans were also keenly aware of the power of portrait painting, commissioning paintings of their founder that bring to life the detailed description of Francis's physical appearance recorded by Thomas of Celano so that a follower could conjure up a mental image of their founder at any time, visualizing his head of

> moderate size and round; his face somewhat long and prominent, his forehead smooth and small; his eyes were black, of moderate size, and with a candid look; his hair was dark, his eyebrows straight; his nose symmetrical, thin, and straight: his ears upright, but small; his temples smooth … His teeth were set close together, white and even; his lips thin and

fine, his beard black and rather scanty, his neck slender; his shoulders straight, his arms short, his hands attenuated, with long fingers and nails; his legs slight, his feet small, his skin fine, and his flesh very spare.[67]

Further evidence of the Franciscan interest in harnessing the powers of the sense of sight can be found throughout other paintings in Assisi. In addition to Pietro's illusionistic glass liturgical vessels and the mirror in the *Allegory of Obedience,* Andrea dei Bartoli painted a series of frescoes in 1367 in the Lower Church, which included a pair of eyeglasses as well as a magnifying lens within the scene of the *Philosophers Confronting Saint Catherine.* A few decades later the Franciscans in Padua commissioned Giusto de' Menabuoi to paint the Chapel of the Beatified Luca Belludi in the Basilica of Saint Anthony where, in the scene of the *Miracle of the Tower,* the artist yet again included a depiction of early modern spectacles.[68]

While the Franciscans seem to have been particularly interested in optical issues, it should be noted that they were not the only order receptive to these matters. In fact, the earliest known depiction of eyeglasses was made for the Dominican Order. In 1352, Tommaso da Modena, who is also responsible for a reliquary with transparent glass discussed in more detail in Chapter 6, painted a several large frescoes for the Dominicans at San Nicolò in Treviso, which include depictions of various optical devices.[69] Cardinal Nicholas of Rouen (d. 1325) is featured with a magnifying glass, the preacher Pietro Isnardo da Chiampo of Vicenza (d. 1244) has a concave reading mirror on his shelf, Saint Jerome (ca. 341–420) is shown with a reading mirror in a leather case resting on the shelf, and Cardinal Hugh of St. Cher (ca. 1200–63) is shown using an early form of eyeglasses.[70]

Looking at the depiction of Hugh of St. Cher using the eyeglasses, it is not hard to image that he is composing his commentaries on *The Sentences of Saint Peter Lombard* while the original source material sits propped open on his shelf. If so, this might add another layer of meaning to the scene, as *The Sentences* were an important source for the church's understanding of the role of vision. Cardinal Hugh's commentary on this text was very influential and one of the earliest to exhibit the influence of Aristotle.[71]

Regardless of his specific activity, the format of eyeglasses he uses is worth further consideration. Unlike modern eyeglasses but typical for the time, those worn by Cardinal Hugh of St. Cher do not have arms extending from the lenses to the ears. Positioned so close to the eyes, these pieces of round glass function almost like external extensions of the sense organ, gathering more information, an interpretation made stronger if one was aware of the glass-like circular tunics of the eye's anatomic structure discussed in the last chapter. This design, with such emphasis on the circular shape of the glass, also creates a strong visual correlation between the spectacles and the prominent round shape of the relic windows in the reliquary panels by artists such as Naddo Ceccarelli, Bartolo di Fredi, and Pietro Lorenzetti.

It is worthwhile to contextualize the reliquaries and their relic windows within the optical climate by considering them in relation to the invention and dissemination of optical technologies made from glass because, by the time viewers were using reliquaries, artists and churchmen alike were familiar with eyeglasses and other visual aides. Spectacles were invented around 1285 in or near Pisa and, within about two decades, the Florentine public knew of their invention.[72] Characteristic of a time when science and technology where intimately connected to religious pursuits, news of the invention reached Florence by way of a Lenten sermon given by Fra Giordano in 1306.[73] As the chronicler reported,

> It is not yet twenty years since there was found the art of making eyeglasses, which make for good vision, one of the best arts and most necessary that the world has. And it is so short a time that this new art, never before extant, was discovered. And the lecturer [Giordano] said: I saw the one who first discovered and practiced it, and I talked to him.[74]

Vincent Ilardi notes that Giordano's sermons that were recorded in the vernacular numbered roughly seven hundred and were widely disseminated, bringing news of the invention to a large segment of the population.[75] It might even be possible that Giotto learned of them around this time.[76]

The spread of the new invention was facilitated by the friar Alexander della Spina, who fashioned an early pair of spectacles and widely discussed his method for doing so. In the *Ancient Chronicle of the Dominican Monastery of Saint Catherine in Pisa*, the friar Bartolomeo da San Concordio (d. 1347) recorded that Alexander della Spina was,

> a modest and good man, whatever he saw that had been made, he knew how to make it. Eyeglasses, having first been made by someone else, who was unwilling to share them, he [Spina] made them and shared them with everyone with a cheerful and willing heart.[77]

Thus, even if the initial inventor kept his craft secret as the chronicler suggests, because of industrious individuals like Spina, the use of eyeglasses spread quickly.[78]

Knowledge and manufacture of eyeglasses had reached a vigorous pace, especially in Venice and Florence, during the fourteenth and fifteenth centuries.[79] By 1321, Venice was already exporting such a high volume of eyeglasses that they instituted a five percent export duty.[80] Orders placed for various types of glasses suggest that Florence had become a thriving center of production during the following century, when the Duke of Milan wrote to his Florentine ambassador saying:

> Because there are many who ask us for eyeglasses that are made there in Florence, since it is reputed that they are made more perfectly [there] than at any other place in Italy, we wish and charge you to send us three dozens of the aforesaid eyeglasses placed in cases so they will not break; namely, one dozen of those apt and suitable for distant vision, that is for the young; another [dozen] that are suitable for near vision, that is for the elderly; and the third [dozen] for [more] common vision.[81]

As trecento viewers encountered or used them, eyeglasses would have transformed one's visual landscape. Older individuals could now work longer, employing their vast accumulated knowledge long into their lives. This would have been particularly useful for senior theologians and natural scientists, but was also helpful for more domestic or artistic endeavors. When writing his 1289 treatise on how to properly take care of the household Sandro di Popozo recounts how, "I am so debilitated by age that without the glasses known as spectacles, I would no longer be able to read and write."[82] As his testimony indicates, individuals from all levels of society were coming to rely on their sense of sight and new types of optical technology.

Having established the many ways round panels of glass—primarily as eyeglasses or mirrors—may have influenced society on a more practical level, it is worthwhile to return the discussion to the relic windows of the reliquaries with which it began. As this chapter argues, the strong formal characteristics shared by the optical devices and relic windows may have forged a strong association between the two, but, there were other factors aside from visual similarities that may have also contributed. In both optical devices and relic windows, the glass functions in a manner to augment or enhance one's earthly sight. In the case of eyeglass, the medium of glass enabled one to see farther. In the case of the reliquaries, glass focused one's gaze on the rather unsightly relics as a way to inspire devotional insight.

To further consider the specific devotional context of the relic windows, it is important to address the fact that many of the reliquaries listed above date to the second half of the fourteenth century and several were made in and around Siena, making it necessary to first situate these objects and their devotional meaning within the context of the debate about the nature of post-plague culture.

The Black Death and Windows to Another World

The outbreak of the bubonic plague in 1348, referred to as the Black Death, was not the only bout of the deadly disease during the early modern period but it has become an important cultural marker because of the devastating toll it took. Siena was among the hardest cities hit, losing approximating three quarters of its population, with the Lorenzetti brothers thought to be among the casualties.

Since the middle of the twentieth century, much of the scholarship about fourteenth-century Italian art has been divided into pre-plague and post-plague art, in accordance with Millard Meiss's thesis that the effects of the Black Death resulted in reactionary artistic traditions after mid-century. Meiss argues that artists abandoned the naturalistic style found in the early part of the century, almost as a form of penance, and returned to more conservative religious themes and styles. No longer did the artworks look to communicate with their viewers on a human level, instead, they referenced the symbolic nature of an abstract, divine realm.

In the aftermath of Meiss, many scholars have questioned and revised his original thesis, contributing to a richer understanding of the complex relationship between the artistic production and the societal changes caused by the plague. One of the most successful recent attempts to offer an alternative model is Judith Steinhoff's book, *Sienese Painting After the Black Death*. Steinhoff astutely maps the evolution of the debate starting with Meiss and traces it through its various scholarly iterations, noting important contributions by Diana Norman, Joanna Cannon, H. W. van Os, and others. For her part, Steinhoff approaches trecento Sienese art as an independent entity, not as a late development in medieval art or as a precursor to the Renaissance. She also rejects the all too common bias for newness and inventiveness instigated by Vasari and dominant in subsequent scholarship. Rather, she presents a compelling case for valuing plurality, for seeing the multiplicity of styles and themes as meaningful choices not mere indications of an artist's abilities.

The group of reliquaries presented here adds an important layer to this conversation for several reasons. Artists began making reliquaries with glass prior to the 1348 plague and they continued production throughout the second half of the century, meaning these objects offer a continuous line of inquiry throughout the entire trecento. Secondly, the painted imagery on some of the reliquaries from the second half of the century is quite naturalistic, conveying three-dimensional forms, individualized physiognomies, and sensitive portrayals of expression. Thus, they challenge Meiss's thesis that artists rejected the realistic style in favor of a more abstract one. The reliquaries, instead, support Steinhoff's pluralistic approach in their variety and juxtaposition of gilded background, naturalistic figures, and transparent glass relic windows.

Though most traditional scholarly accounts do not consider the later fourteenth century a time of particular artistic innovation, I argue that inventiveness can indeed be found within these reliquaries, albeit in a slightly different form.

Rather than the traditional large-scale public naturalistic narrative frescoes that have come to characterize the period, the reliquaries discussed here don't utilize naturalism to foster unity between the actual space and the painting space. Instead, they use naturalistic imagery in combination with other styles and materials in order to offer multiple visual modes and different avenues for contemplating the divine. The reliquaries do not present a cohesive, illusionistic view; instead, they embrace and thrive on the juxtaposition of various styles and materials. These artists were exploring different ways of using visual media to inspire spiritual insights, capitalizing on the newly available medium of glass and the growing availability of relics to create devotionals tools capable of offering powerful moments of hope and comfort to a weary, plague-ridden community. What follows here contextualizes the reliquaries with relic windows within the atmosphere of post-plague Italy in the sense that particular attention is paid to how the reliquaries could help one meditate on the nature of the heavenly afterlife by using the relic windows as something akin to imaginary spiritual portals, a concept explained in more detail shortly.

To do this, it is necessary to again return to the shape of the relic windows. Both the circular shaped windows, as seen examples such as that Naddo Ceccarelli, and the quatrefoil format found in Lippo Vanni's example in the Vatican Collection, had contemporaneous references that would have reinforced their role as dynamic spiritual portals. The quatrefoil shape enjoyed great popularity throughout the Gothic period in a range of media, from sculptural detailing to painted border decorations in frescoes and panel paintings to actual windows. A common place to find actual windows of this shape is in the tracery above the mullion, as seen in the background of the illusionistic niche by Giotto in the Arena Chapel (Figure 4.6) already discussed in Chapter 4 for its depiction of glass lamps.

The circle shape was also used widely in early modern art; it also had a plethora of distinguished associations dating back to antiquity. It was referenced extensively in discussions of the celestial spheres and cosmology, in the work of contemporary literature like that of Dante and Saint Catherine of Siena, valued by the ancient Greeks and Neoplatonists for its geometric purity, and its round form was even cited in the *Speculum humanae salvationis* in relation to Mary's immaculate nature due to its lack of "untidy corners."[83] Another important association with the circle is found in the many monstrances, or ostensoria, which often displayed the host—a type of relic—through a round piece of glass or crystal as in the case of the *Ostensorium with the Paten of Saint Bernward* in the Cleveland Museum of Art (inventory no. 1930.505).

The most prominent example of circular windows during the trecento was undoubtedly the rose window featured in many Gothic churches throughout Italy and the rest of Europe. Considering that most of the reliquaries with circular relic windows come from Siena, it is worth recalling that the famous rose window in that city's cathedral was designed by Duccio in the late thirteenth century discussed in Chapter 2. Duccio's oculus would have dominated the façade of the town's most important building and, as the term oculus suggests, it may have been viewed as a reminder of the watchful divine eye surveying and protecting Siena. Though it is made of stained glass and the relic windows are colorless glass, the comparison is still worth considering because when Duccio's window was viewed from the outside, the color of the glass did not register as much as the impact of its circular shape.

There were also less lavish—but perhaps more relevant—examples of round windows. As with the case with quatrefoil windows, fourteenth-century paintings with scenes of everyday life by Pietro Lorenzetti and others make several references to circular windows. The townscape featured in Pietro's famous altarpiece depicting the *Birth of the Virgin* altarpiece for the cathedral of Siena includes round windows, as does his brother's *The Effects of Good Government in the City* in the Palazzo Pubblico of Siena.[84] Making this connection even stronger is the fact that when windows were glazed, they were sometimes filled with small round transparent glass pieces—sometimes referred to as crown glass or "bullseye" glass—as seen in the case of the round windows in the chapel of the Cardinal of Portugal and an illustration of *Petrarch in his Study* from a frontispiece to his *Viris illustribus* (Figure 5.7).[85] Perhaps not coincidentally this is also the type of window glass treatment found throughout the Arena Chapel's side windows, choir windows, and windows on the façade. The repeating shape of these circular forms within the glass windows could have either fostered, reinforced, or reflected a potential association between actual windows and the relic windows.

If we expand the discussion to include depictions of other types of apertures—specifically those of a more spiritual nature—even more fruitful lines of investigation come to light. There are important paintings of illusionistic openings shaped like circles and quatrefoils made by the very same artists mentioned throughout this chapter. The ceiling of Giotto's Arena Chapel is filled with golden stars set against a deep blue sky; interrupting this continuous span of night sky are circular apertures that function like skylights, allowing the viewer to peek through the sky into a heavenly space above (Figure 5.8). These portals to heaven reveal divine figures peering down from their golden celestial realm. The holy figures clearly belong

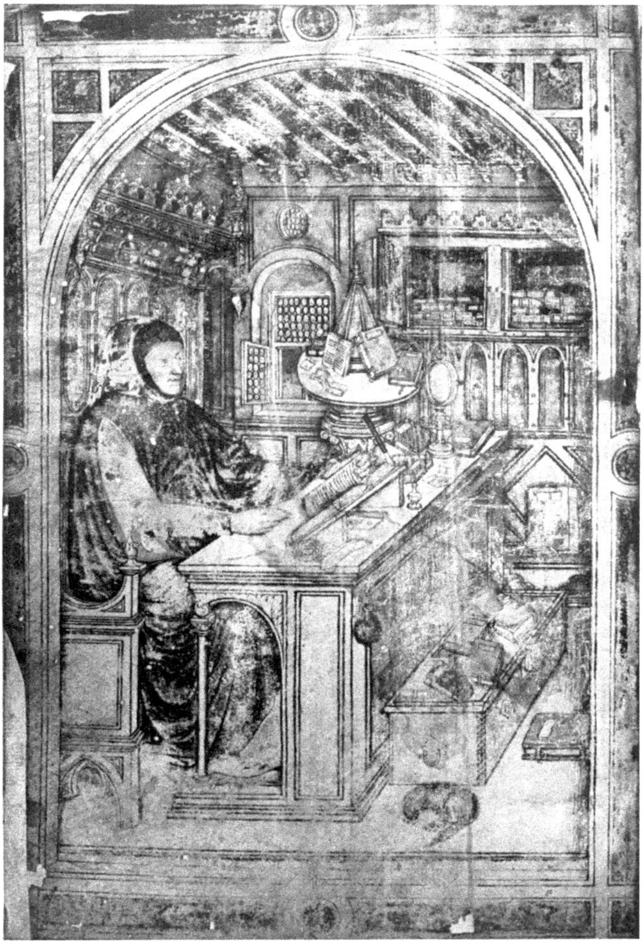

Figure 5.7: Unknown Artist in the Style of Altichiero, *Petrarch in His Study*, from the frontispiece of Petrarch's *De viris illustribus,* ca. 1400, Darmstadt, Universität- und Landesbibliothek, MS 101. Source: Public Domain via Wikimedia.

to another reality, and yet, at the same time, they seem to be able to communicate with the visitor below, making eye contact or gestures that suggest the possibility of a connection between these two worlds.

The same holds true when considering the quatrefoil shaped relic windows found in the reliquaries by Lippo Vanni in the Vatican Collection and Simone dei Crocifissi's in a private collection. In the ceiling of the Baroncelli Chapel in Santa Croce (Figure 5.9), Taddeo Gaddi painted illusionistic openings similar to those by Giotto. In each of the four segments of the vault overhead, Taddeo depicted

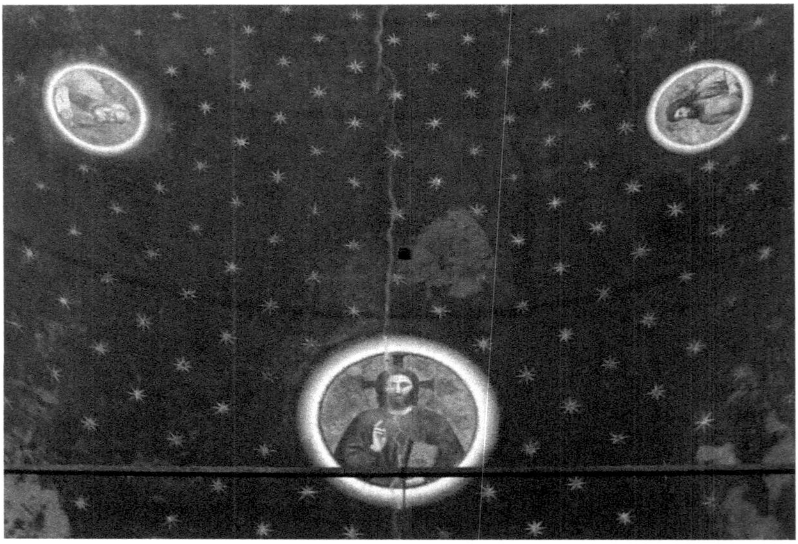

Figure 5.8: Giotto, Detail of Arena Chapel Ceiling, ca. 1305, Padua. Source: José Luiz Bernardes Ribeiro via Wikimedia Commons (CC BY-SA 4.0).

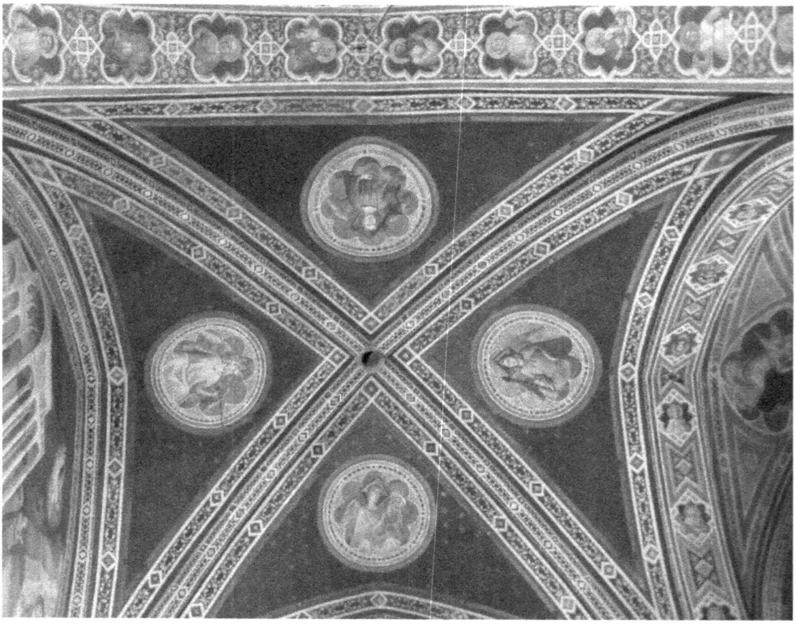

Figure 5.9: Taddeo Gaddi, Detail of Baroncelli Chapel Ceiling, ca. 1330, Santa Croce, Florence. Source: Francesco Bini via Wikimedia Commons (CC BY 2.5).

a circular portal framed by scallop-shaped decorations with divine figures set against a golden background peering down into the space below.

The vaults, ceilings, and domes of such religious spaces were oftentimes equated with heaven while the frescoed scenes on the walls tended to show narrative scenes filled with naturalistic figures in three-dimensional settings and backgrounds filled with blue skies. Thus, characteristic of their time, the two chapels seem to reference the mundane realm on the walls and the heavenly one in the vault. This arrangement is common and logical, as the viewers' eyes ascended to the uppermost parts of the church, their mind was also uplifted to thoughts of the heavenly sphere. When looking up toward the ceilings of the Arena Chapel and Baroncelli Chapel, viewers would have found confirmation of the spiritual realm looming just beyond the physical reality because, visible through the illusionistic windows puncturing the ceilings, were glimpses of the promise of heavenly salvation. The viewers were able to peek into celestial sphere, represented by its golden glow and divine figures. Thus, while the majority of the imagery on the chapel walls references the material world through naturalistic imagery and stories that took place on earth, there are small moments alluding to the presence of a simultaneous plane of the spiritual world.[86]

The reliquaries offered a similar convergence of heavenly and earthly, however, they operated in the reverse manner. The central panels of the reliquaries discussed here primarily present a spiritual scene. In the case of Naddo Ceccarelli's reliquary, the central scene features an elegant Madonna with her child standing before an ambiguous golden background. She is not an earthly mother but rather a heavenly one. The panels' overwhelming use of gold leaf, the pointed gables with ethereal flame-like finials, and the apparition-like quality of the figures simulate an experience of divine vision. The relic windows literally puncture this golden framework to assert that the rocks, bones, and material objects of the saints have a place in, or at least a connection to, heaven. These relic windows are reminders that the saints, the Virgin Mary, and Christ participated in the earthly, bodily, physical reality at some point.

The reliquaries' dominant visual mode, on the other hand, which seems more catered to an abstract manner of contemplating the divine, is intermittingly interrupted by the presence of relic windows and relics in the frames surrounding the central scene. Thus, in the reliquary the majority of the imagery is related to a spiritual setting yet it also features brief allusions to mundane reality whereas the chapels do the reverse, they primarily feature human dramas in realistic paintings with infrequent glimpses into a more ethereal setting through the illusionistic windows overhead. Both the fresco cycles and relic windows seem to reference moments of convergence between the mundane and spiritual worlds. The two realms are united, they work together, revealing and concealing, appearing and disappearing at various times.

Reliquaries capable of presenting this complex, interconnected relationship of the two worlds would have been particularly appreciated in post-plague Siena because the high mortality rate likely inspired a pervasive concern about one's relationship to their afterlife and the spiritual realm. Reliquaries in general, but the ones discussed here in particular, bring relics to life by setting them within their theological context. Like the Eucharist, these bones, stones, and cloth were tangible loci of the intersection between heaven and earth. These reliquaries reinforced the unity of matter and spirit by celebrating relics, fragments of deceased bodily remains, alongside a depiction of divine figures such as Mary or Christ. Such an idea may have been comforting in the context of a post-plague society, or any society in which death was so present. The boundary separating the living and the dead seemed permeable and the reliquaries were dynamic portals through which to commune with another reality. Like a church, these objects brought one to the threshold of the divine.

However, while relics were certainly some of the most inspirational and venerated objects possessed by the Church, one must acknowledge their limitations, the primary of which is their intercessory nature. Mary and Christ left no primary, that is, bodily, relics. The closest one could get was something worn by them or touched by them. Therefore, like the saints themselves, relics could not offer a direct union with God the Father. This theory might explain the placement of the relic windows along the periphery of the panels. By placing the relics on the margins and putting focus on imagery featuring Christ and the Virgin Mary, the theological emphasis matched the compositional one. In other words, in terms of church hierarchy, the saints were intercessors, and correspondingly, the reliquaries present imagery related to Christ and Mary in the center surrounded by the relic windows.

In order to fully contemplate the most sacred mysteries of the Christian faith one needed to mentally ascend from contemplation of the intersection of the heavenly and earthly to direct meditation on God. For direct knowledge of the God, however, one had to wait until after death for the moment of the Beatific Vision. Devotional art such as reliquary panels could help one contemplate this union with the divine in powerful ways, and this was especially true in the case of the third subgroup, that is, the reliquaries with verre églomisé glass. These reliquaries will be discussed in more detail in the following chapter because they use a combination of gilded glass and transparent glass, which results in a very different visual experience from the transparent relic windows described above. And, as I will argue, the reliquaries with verre églomisé offer a unique opportunity to contemplation of the Beatific Vision.

Conclusions on Transparent Glass in the West: Seeing Glass through a Renaissance Lens

In summary, with the increasing availability of clear glass, many reliquaries begin to incorporate the medium in order to meet the congregation's desire to directly view relics and adhere to the mandate that relics also be contained. At the same time, theologians capitalized on vision as a spiritual metaphor, poets referenced the transformative power of visual experiences, and the market catered to the increasing demand for high quality lenses and eyeglasses. When one pulls together all the various threads related to optics and glass, it seems that transparent glass indeed functioned as a thinking tool, to borrow the phrase of Macfarlane and Martin. Emerging glass technologies and artistic applications inspired new ideas about vision and shaped the early modern subject's relationship with the Divine in possibly profound ways.

Glass, in its various forms, provided trecento artists with a multifunctional medium that could inspire a variety of complex devotional strategies. The use of transparent glass for relic windows resonated with the growing interest in the sense of sight and the value of empirical observation. Using one's powers of observation, which had been honed to more careful looking due in part to the new optical technology and interest in the natural world, one could study the realistic representations of the holy figures and appreciate their lifelike qualities. Just as Hugh of St. Cher peered through his eyeglasses to study or write the text in front of him, the glass relic windows framed and focused one's gaze on the spiritual potency of the relics. In this respect, the use of glass evokes the ideas of Aristotle in two ways: first, the glass lenses facilitated detailed empirical study of matter as a means of obtaining information and, second, glass panels resonated with the notion of the intervening medium, an element crucial for establishing a visual connection within his optical theory. Thus, the use of transparent glass in artworks seems to correlate with the growing interest in Aristotelian thought and access to visual aides. Such a theory could certainly help explain the growing interest in the natural world as a whole and, in turn, in naturalistic representations in art. However, as the next chapter will show, the situation was not that straightforward. The increasing influence of Aristotle and the rise of naturalism are just one part of a more complex whole. In fact, the ascendancy of Aristotle's visual theory did not completely negate that of Plato. Rather, the two existed simultaneously. The next chapter considers how one might navigate competing visual theories within the context of devotional as it analyzes reliquaries that incorporate a complex blend of both gilded glass and transparent glass.

Notes

1. Alan Macfarlane and Gerry Martin, *Glass: A World History* (Chicago: University of Chicago Press, 2002), 43. As Macfarlane and Martin, *Glass*, 43–45 notes

   > It could also be argued that windows altered thought at a deeper level. The question here is the way in which glass, whether in a mirror, window, or through a lens, tends to concentrate and frame thought by bounding vision, and at the same time leads to abstraction and attention to the details of nature. It seems likely that the glass window altered the relations between humans and their world in ways which it is now difficult to recover. It may have encouraged the contemplation of external nature from within the house, an appreciation of nature for its own sake, seen through a window.

   However, Macfarlane and Martin, *Glass*, 47–48, also acknowledge that

   > obviously glass, on its own, is not enough. Without the burst of curiosity and new knowledge from ancient Asian civilizations, all the glass in the world would probably have had little influence on thought. It is the combination of curiosity and tools that is important. … Yet glass, it seems to us, is a *sine qua non* of the development of the experimental method we call science.

2. Macfarlane and Martin, *Glass,* 14. For additional commentary on this idea, see ibid., 3, 40, 51–59. For similar arguments related to the relationship among glassmaking, technology, natural philosophy, and the study of matter in the ancient world see Marco Beretta, *The Alchemy of Glass: Counterfeit, Imitation, and Transmutation in Ancient Glassmaking* (Sagamore Beach, MA: Science History Publications, 2009).
3. Vincent Ilardi, *Renaissance Vision from Spectacles to Telescopes* (Philadelphia: American Philosophical Society, 2007), 27–28.
4. As Timothy J. Johnson, "Francis and Creation," in *The Cambridge Companion to Francis of Assisi,* ed. Michael J. P. Robson (Cambridge: Cambridge University Press, 2012), 144–45 explains, Francis was "free to embrace the reality of the material world and enter into a relationship with creation on its own terms and praise the Creator." For the distrust of the senses in the earlier medieval period see Richard G. Newhauser,

"Peter of Limoges, Optics, and the Science of the Senses," *Sense and Society* 5, no. 1 (2010): 29. For more on the shift to a more visually inclined culture, see Alfred W. Crosby, *The Measure of Reality: Quantification and Western Society, 1250–1600* (Cambridge: Cambridge University Press, 1997), 3–19.

5. For more on the optical theories reflected in the artworks mentioned see Hayden Maginnis, *The World of the Early Sienese Painter* (University Park, PA: Pennsylvania State University Press, 2001), 178–84. For more on Taddeo Gaddi's exposure to a solar eclipse, see chapter 2, note 42.
6. Illustrated in Maginnis, *World of the Early Sienese Painter*, fig. 101.
7. See Eliot W. Rowlands, "Sienese Painted Reliquaries of the Trecento: Their Format and Meaning," *Konsthistorisk Tidskrift* 48 (1979): 122–38 and Virginia Brilliant, "A Framework for Devotion in Trecento Siena: A Reliquary Frame in the Cleveland Museum of Art," *Peregrinations: Journal of Medieval Art and* Architecture 4, no. 3 (Spring 2014): 66–94 for a discussion of the type of reliquary under discussion in this chapter. This study both expands the examples in these publications and focuses more on an analysis of transparent glass roundels within a network of glass.
8. There is a fourth group of reliquaries which deserves brief mention because they feature a device similar to relic windows in the frame surrounding a central devotional image but, because they date to the early fifteenth century, are beyond the scope of this book. Sometime around 1424–34 Fra Angelico painted four reliquary tabernacles, two of which feature apertures surrounding the central panel (*Annunciation and Adoration of the Magi* and *Madonna delle Stelle*). For more information on these, see the forthcoming exhibition catalog from "Fra Angelico: Heaven on Earth" held at the Isabella Stewart Gardner Museum in Boston. Though the small openings in these reliquary tabernacles do not emphasize the circular shape they are still worth considering as they may have been related to the quatrefoil-shaped openings in some of the reliquaries discussed here. The ways in which Fra Angelico's reliquaries resonate with a network of glass and other optical issues requires further attention, however, it is worth noting that Fra Angelico painted highly detailed transparent glass oil lamps in a scene depicting a vision to Saint Lucy, the patron saint of sight. The panel, *Saint Agatha Arising from her Tomb and Appearing to Saint Lucy and her Mother Eustachia*, was formerly in the Feigen Collection and is now in a private collection. For more, see Laurence Kanter, "A Rediscovered Panel by Fra Angelico," *Paragone* 599 (January 2000): 3–13. I would like to thank Christopher Platts for drawing my attention to this depiction of glass lamps.
9. See Rowlands, "Sienese Painted Reliquaries," 123; C. Griffith Mann, "Relics, Reliquaries, and the Limitations of Trecento Painting: Naddo Ceccarelli's Reliquary Tabernacle in the Walters Art Museum," *Word and Image* 22, no. 3 (July–Sept. 2006): 251–59; and C. Griffith Mann, "Naddo Ceccarelli, Reliquary Tabernacle with Virgin and Child," in *The Walters Art Gallery: Guide to the Collection* (London: Scala Books, 1997), 24.
10. C. Griffith Mann's catalog entry on this object (Mann, "Naddo Ceccarelli," 24), identifies the relics as follows:

(inner frame, clockwise from top) illegible (pair), illegible, a bone of Saint John the Baptist, a stone from the column of the flagellation, illegible, a stone from Calvary, a stone from … [illegible], a stone from the place where Christ's cross was found, illegible, a nail from the gates of Jerusalem (?), a relic from the sepulcher of … [illegible], illegible, illegible, a stone from the sepulcher of Saint John the Evangelist, a stone from the ground where the wood of the cross stood (?), a stone associated with Mary Magdalene, a piece of the stone from which Christ ascended, illegible, a stone from the sepulcher of Christ, illegible, a stone from the sepulcher of Christ; (left pilaster from top down) bones from Saint Barnabas the Apostle, a fragment of the tunic of Saint Francis; (right pilaster from top down) illegible, illegible, a piece of the tunic of Saint Agnes (?), a relic of Saint James the Apostle, a relic of the sepulcher; (rectangular chamber from left to right) the bones of Saint Peter Martyr, relic of Saint Damascus Pope, the bones of Saint Gregory the Great, a relic of Saint Alexis, a stone from the temple of God, stones used in the martyrdom of Saint Stephen, an illegible saint's relic, and a relic of the True Cross (?).

11. Rowlands, "Sienese Painted Reliquaries," 128. This reliquary is illustrated at ibid., 129.
12. Mann, "Relics," 252.
13. C. Griffith Mann, "Reliquary Triptych with the Annunciation, Saint Ansanus, the Adoration of the Magi, and the Crucifixion," in *Treasures of Heaven: Saints, Relics, and Devotion in Medieval Europe*, eds. Martina Bagnoli and others (New Haven, CT: Yale University Press, 2010), 205. As Mann observed here, "This program takes its point of departure from the wings, which introduce the theme of spirit's entry into matter. Combined with episodes from the beginning and the end of Christ's earthly life, the relics extended the theme of the Incarnation into the core of the object. Moreover, the move from painted to sculpted decoration reinforced the shift from spiritual to physical presence."
14. Arnold Angenendt, "Relics and Their Veneration," in *Treasures of Heaven: Saints, Relics, and Devotion in Medieval Europe* (New Haven, CT: Yale University Press, 2010), 23. As Angenendt describes it, "a veritable pilgrimage fever broke out" because after tenth century, it was thought that the saint's body could be divided up and each relic fragment was just as sacred as the entire complete whole.
15. Rowlands, "Sienese Painted Reliquaries," 124.
16. Ibid., 125.
17. Ibid., 125 n. 25, "Ne reliquiae sanctorum ostendatur extra capsam; ne novae habeantur in veneratione sine Romana ecclesia." Transcribed in Ernst Günther Grimme, *Goldschmiedekunst im Mittelalter: Form und Bedeutung des Reliquars von 800 bis 1500* (Cologne: DuMont, 1972), 164.
18. Ilardi, *Renaissance Vision,* 10.
19. For more on reliquaries as houses see Angenendt, "Relics and Their Veneration," 25.
20. As the catalog of relics from Mann, "Naddo Ceccarelli," 24 indicates, the group of relics contains a large group of stones and bones.

21. This reliquary is illustrated (fig. 121) and discussed in Martina Bagnoli and others, *Treasures of Heaven: Saints, Relics, and Devotion in Medieval Europe* (New Haven, CT: Yale University Press, 2010), 204–5.
22. These three pinnacles could also be a reference to the three pinnacles of Solomon's Temple as noted by Edward Olszewski, "A Possible Source for the Triptych, Lunette, and Tondo Formats in Renaissance Paintings," *Notes in the History of Art* 28, no. 2 (Winter 2009): 5.
23. For more on Giotto's *Annunciation* and the way it transformed a Roman triumphal arch into a gateway to heaven, see Jules Lubbock, *Storytelling in Christian Art from Giotto to Donatello* (New Haven, CT: Yale University Press, 2006), 42.
24. Diana Norman, "Those Who Pay, Those Who Pray and Those Who Paint: Two Funerary Chapels," in her *Siena, Florence, and Padua: Art, Society and Religion: 1280–1400,* vol. 2 (New Haven, CT: Yale University Press, 1995), 171.
25. Bartolo di Fredi's *Diptych with Annunciation* from ca. 1383 in the Szépművészeti Múzeum, Budapest is one example. A second is an altarpiece in Montalcino, which Bartolo painted in the 1380s for the shoemaker's guild. This large polyptych featured *The Coronation of the Virgin* and was displayed in conjunction with two large polychrome statues of the Virgin Annunciate and the Angel Gabriel by Angelo di Nalduccio from ca. 1370. It is thought that these sculptures flanked the central panel. See Catherine King, "The Arts of Carving and Casting," in *Siena, Florence, and Padua: Art, Society and Religion: 1280–1400*, ed. Diana Norman, vol. 1 (New Haven, CT: Yale University Press, 1995), 116–17.
26. For references to the increasing visibility of relics see Holger Klein, "Eastern Objects and Western Desires: Relics and Reliquaries between Byzantium and the West," *Dumbarton Oaks Papers* 58 (2004): 305 n. 125 where he remarks, "In making the relics visible, the Trier artists followed a trend attested in western art from the later 12th and early 13th century onward." For a similar argument see Arnold Angenendt, "Relics and Their Veneration," in *Treasures of Heaven: Saints, Relics, and Devotion in Medieval Europe* (New Haven, CT: Yale University Press, 2010): 25–26 and Caroline Walker Bynum, *The Resurrection of the Body in Western Christianity 200–1336* (New York: Columbia University Press, 1995), 202–3 which notes,

> After 1150, however, what German historians call expressive or "speaking" reliquaries became popular—reliquaries … that indicate by their form the nature of the fragment. In the twelfth century, such containers were still sheaths of gold and jewels, which revealed fragmentation but masked decay. By the thirteenth century, they began to contain windows of crystal through which shards of tibia or bits of finger could be viewed. The same period saw the emergence of ostensoria—containers, parallel to and sometimes identical with Eucharistic monstrances, that were made especially for displaying bits of holy bodies or even the fluids they produced before or in death.

For more on the issue of the visible relic in general see Hans Belting, *Likeness and Presence: A History of the Image Before the Era of Art* (Chicago: University of Chicago Press, 1994), 303; Martina Bagnoli, "The Stuff of Heaven: Materials and Craftsmanship in Medieval Reliquaries," in *Treasures of Heaven: Saints, Relics, and Devotion in Medieval Europe*, eds. Martina Bagnoli and others (New Haven, CT: Yale University Press, 2010), 141–42; Stefania Gerevini, "*Sicut crystallus quando set objecta soli*: Rock Crystal, Transparency, and the Franciscan Order," *Mitteilungen des Kunsthistorischen Institutes in Florenz*, 56, no. 3 (2014): 255–83; and Stefania Gerevini, "*Christus Crystallus*: Rock Crystal, Theology, and Materiality in the Medieval West," in *Matter of Faith: An Interdisciplinary Study of Relics and Relic Veneration in the Medieval Period,* eds. James Robinson and Lloyd de Beer with Anna Harnden (London: British Museum Press, 2014) 92–99. Although the work of Stefania Gerevini is focused on the use of rock crystal rather than glass it is still important for this discussion because of the many related themes of transparency.

27. Klein, "Eastern Objects and Western Desires," 283–314 and Bagnoli, "The Stuff of Heaven," 140–41.
28. For an extensive list of relic exchanges with literary documentation see Holger Klein, "Eastern Objects and Western Desires," 284–93. Two of the more famous episodes of Western visitors to the Byzantine court included Henry the Lion and King Louis VII of France, who visited the Emperor Manuel I Komnenos on their journeys to the Holy Land. At the Byzantine court, Louis VII and Henry would have hoped to see such revered relics as the column and whip of Christ's flagellation, the wood and nails from the True Cross, with the emperor's permission. Ibid., 287.
29. Examples include the *Cup with Mount* and *the Rock Crystal Vase with Mount* in the Treasury of San Marco, Venice. Stefano Carboni, *Venice and the Islamic World: 828–1797* (New York: Metropolitan Museum of Art, 2007), 19. For more on the shipload of tesserae that entered Venice, see Eve Borsook, *Medieval Mosaics: Light, Color, Materials* (Milan: Silvana, 2000), 13. For more on the many pieces of Islamic luxury arts in the Byzantine treasury, see Shalem, *Islam Christianized,* 45.
30. For Holger Klein, the Byzantine influence on European reliquary production is already observable in the late twelfth century. Klein, "Eastern Objects and Western Desires," 299. Ibid., 296 notes, "the arrival of Byzantine relics in the West is only rarely attested during the eleventh century." See also Cynthia J. Hahn, *Strange Beauty Issues in the Making and Meaning of Reliquaries* (University Park, PA: They Pennsylvania State University Press, 2012), 231.
31. For more on how this object resonates with this chapter's thesis, see Bagnoli, "The Stuff of Heaven," 141–42, 176–77 where she describes the process of transformation from a gilded container to a panel with visible relics through translucent horn. Ibid., 142 finds this particular case an illustration of the "movement of the relics from the hidden core of the reliquary to the periphery" and a testament to "the shift from materiality to visuality that occurred in thirteenth-century art. Where the gems had

once stood metaphorically for the bones of the saints, they now yielded to the actual remains of those saints."

32. Macfarlane, *Glass,* 14.
33. The Metropolitan Museum of Art's collection features examples such as inventory nos. 91.1.1230, 81.10.3a,b; and 20.254a,b.
34. C. Louis Avery discussed the traditional theory that they were used for identification purposes as well as Eisen's theory that they served a protective capacity, warning evil spirits to stay away from the Christians. For more on this debate and for further bibliography see C. Louis Avery, "Early Christian Gold Glass," *The Metropolitan Museum of Art Bulletin* 16, no. 8 (August 1921): 170.
35. Gary Dickson, "The Crowd at the Feet of Pope Boniface VIII: Pilgrimage, Crusade and the first Roman Jubilee (1300)," *Journal of Medieval History* 25, no. 4 (1999): 280–81.
36. John Huizinga, *The Waning of the Middle Ages* (Garden City, NY: Doubleday Anchor, 1954), 284.
37. Alfred W. Crosby, *The Measure of Reality: Quantification and Western Society, 1250–1600* (Cambridge: Cambridge University Press, 1997). On page 133, Crosby acutely notes that, "A society in which the chief conduit of authority was the ear, tilted to the recitation of Scripture and the church fathers, to the somniferous repetition of myths and epics, began to become a society in which the recipient of light ruled: the eye."
38. Maginnis, *World of the Early Sienese Painter,* 174.
39. Ibid., 175.
40. Roland Recht, *Believing and Seeing: The Art of Gothic Cathedrals* (Chicago: University of Chicago Press, 2008), 71.
41. Ibid., 70–71; Vincent Lorne Kennedy, "The Moment of Consecration and the Elevation of the Host," *Mediaeval Studies* 6 (1944): 121–50; and Miri Rubin, *Corpus Christi: The Eucharist in Late Medieval Culture* (Cambridge: Cambridge University Press, 1991), 55.
42. Recht, *Believing and Seeing,* 70–73, 94. For more on the visual properties of rock crystal and associations with issues related to the sense of sight see Gerevini, "Sicut Crystallus," 255–83 and Gerevini, "Christus Crystallus," 92–99.
43. For the influence of Aristotle (and others) on Alexander, see Christopher M. Cullen, "Alexander of Hales," in *A Companion to Philosophy in the Middle Ages,* eds. Jorge E. Gracia and Timothy B. Noone (Oxford: Blackwell Publishing, 2005), 104–8.
44. Recht, *Believing and Seeing,* 70–71.
45. Quoted in Maginnis, *World of the Early Sienese Painter,* 181.
46. Roger Bacon writes, "For Aristotle says in the first book of Metaphysics that vision alone reveals the differences of things; since by means of it we search out experimental knowledge of all things that are in the heavens and in the earth. … Therefore it [optics] is the flower of the whole of philosophy and through it, and not without it,

can the other sciences be known." Quoted in Maginnis, *World of the Early Sienese Painter,* 174.

47. David C. Lindberg, *Theories of Vision From Al-Kindi to Kepler* (Chicago: University of Chicago Press, 1976), 209–23 and Maginnis, *World of the Early Sienese Painter,* 176.
48. However, it should be noted that Plato's visual theory continued to receive attention through the fifteenth century when aspects of it were mentioned by both Alberti and Leonardo da Vinci.
49. Lindberg, *Theories of Vision*, 92.
50. It should be noted that early departures from the mostly Platonic tradition were made by William of Conches (ca.1080–1154), Adelard of Bath (fl. 1116–1142), and Robert Grosseteste (c. 1168–1253), however, these theorists still considered themselves proponents of Plato. Lindberg, *Theories of Vision*, 91–102.
51. Lindberg, *Theories of Vision*, 116 for commentary on Aristotle's theories superseding those of Plato. For Albertus's contribution and his attempt to synthesize all previous knowledge on optics, except, that is, for only limited exposure to Alhazen, see ibid., 104–5. In his works, *De anima* and *De sensu*, Albertus Magnus discussed and dismissed many of the previous theories of vision, including those put forth by Plato, Euclid, and Al-Kindi. Ibid., 104–5.
52. Ibid., 105 notes, "In place of these discredited theories, [Albert] attempts to establish the Aristotelian doctrine that vision is caused by an alteration (*immutatio*) of the transparent medium by the visible object and the propagation of this alteration to the watery substance of the eye."
53. Ibid., 6–9. As Lindberg notes, Aristotle "perceives the absolute necessity of this medium … [which is] diaphanous or transparent, a nature or power found in all bodies, but especially air, water, and certain solid substances."
54. Quoted in ibid., 7.
55. Dante Alighieri, *The Convivio*, ed. and trans. Andrew Frisardi (Cambridge: Cambridge University Press, 2018), 173. This entire passage, which is located at lines 6–16 of Chapter 9 of Book 3, describes Dante's understanding of vision. And, as will be mentioned later, this passage also uses the analogy of a glass mirror backed with lead to help explain the visual process.
56. Though *Convivio* may not have been completed or circulated during his lifetime, the fact that Dante wrote the text in the vernacular indicates that he intended its eventual circulation among a wide public.
57. Carl Brandon Strehlke and Machtelt Brüggen Israëls, eds., *The Bernard and Mary Berenson Collection of European Paintings at I Tatti* (Florence: Villa I Tatti in collaboration with Officina Libraria, 2015), Pl. 53.
58. Though it is very difficult to see, an image of this niche was published by Hayden Maginnis in "Assisi Revisited: Notes on Recent Observations," *The Burlington Magazine* 117 no. 869 (August 1975): 514. The image used here is the author's photograph.

59. Pietro's illusionistic niche in the Lower Church may also be an important consideration for analysis of the illusionistic niche in Taddeo Gaddi's Baroncelli Chapel, Santa Croce, Florence, (Figure 2.8). This connection is discussed in detail in chapter 7.
60. While both glass and metal mirrors were available at this time some of these depictions exhibit the convex shape characteristic of a glass mirror. At least some people understood mirrors were made of glass because Dante describes a mirror as glass backed by lead in chapter 9 of book 3 of his *Convivio*. See Dante *The Convivio*, 173.
61. Eloi Leclercq, *The Canticle of Creatures: Symbols of Union,* trans. Matthew J. O'Connell (Chicago: Franciscan Herald Press, 1978), 4; Fiorenza Bajetto, "Un trentennio (1941–1973) di studi sul Cantico di Frate Sole, Bibliografia ragionata," *L'Italia francescana* 49 (1974): 5–62.
62. The question of a specific brand of "Franciscan optics" is discussed in Ilardi, *Renaissance Vision*, 27.
63. Johnson, "Francis and Creation," 153.
64. Leading scientists in other fields were also associated with the church because, as noted by Macfarlane, *Glass,* 45 and Ilardi, *Renaissance Vision*, 27, one of the main reasons the churchmen made such great advancements in science was because they had access to educational resources and time for study. Lindberg, *Theories of Vision*, 107 urges caution in making superficial connections, however. Despite these explanations and cautions it is still important to note the intersections between religion, science, and art and note the fact that Franciscans made notable achievements in each of these fields. Moreover, as will be discussed later in this chapter, many Franciscan paintings depict optical devices and several Franciscan commissions used glass in interesting ways, as will be discussed in chapter 6.
65. Lindberg, *Theories of Vision*, 107. Bacon and Pecham both joined the Franciscan order around 1250 in Oxford. For more on this see David C. Lindberg, "Lines of Influence in Thirteenth-Century Optics: Bacon, Witelo, and Pecham," *Speculum* 46, no. 1 (Jan. 1971): 67. For Duns Scotus, see Katherine H. Tachau, *Vision and Certitude in the Age of Ockham: Optics, Epistemology and the Foundations of Semantics, 1250–1345* (Leiden: E. J. Brill, 1988), 55. For Ockham, see ibid., 113. For Grosseteste's association with the order, see Lindberg, *Theories of Vision*, 107.
66. Maginnis, *World of the Early Sienese Painter,* 175.
67. Thomas of Celano, *Vita prima,* 83; trans. A. G. Ferrers Howell, *The Life of Saint Francis by Thomas de Celano* (London: Methuen, 1908), 81; quoted in Paul Hills, *The Light of Early Italian Painting* (New Haven, CT: Yale University Press, 1987), 12.
68. For other instances of eyeglasses, see Chiara Frugoni, *Medioevo sul naso: Occhiali, bottoni e altre invenzioni medievali* (Rome: Laterza, 2001).
69. For more on Tommaso and his work in Treviso see, Robert Gibbs, *Tomaso da Modena: Painting in Emilia and the March of Treviso, 1340–80* (Cambridge: Cambridge University Press, 1989), 50–87, 257–67.

70. Ibid., and Ilardi, *Renaissance Vision*, 19–21.
71. John Fisher, "Hugh of St. Cher and the Development of Mediaeval Theology," *Speculum* 31, no. 1 (Jan., 1956): 57–69.
72. For an extensive summary on the literature discussing the invention of eyeglasses, see Ilardi, *Renaissance Vision,* 3–49.
73. Ibid., 5.
74. Quoted in ibid., 5.
75. Ibid., 21–22. For more on this, see Daniel R. Lesnick, *Preaching in Medieval Florence: The Social World of Franciscan and Dominican Spirituality* (Athens, GA: University of Georgia Press, 1989), 111.
76. Ilardi, *Renaissance Vision,* 22 notes that among his many travel destinations, Giordano was in Florence from 1302–1305, again from 1306–1307, and also in 1309. According to John White, *Art and Architecture in Italy: 1250–1400* (New Haven, CT: Yale University Press, 1993), 309 Giotto was listed as living in the parish of Santa Maria Novella in 1301. Though Giotto would soon leave to paint the Arena Chapel in Padua, there could have been contact between the two given their close time in Florence, their proximity to Santa Maria Novella, and their shared interest in optics. This potential connection is discussed in more detail in chapter 7.
77. Cited in Ilardi, *Renaissance Vision*, 6.
78. As Ilardi points out in ibid., 7, after observing one pair of glasses, one could easily imitate the basic idea.
79. There is also evidence that Pisa had a thriving glass industry at least from the early years of the fourteenth century and possibly earlier. For more on the glass industry in Pisa see Tito Antoni, "Note sull'arte vetraria a Pisa fra il Tre e il Quattrocento," *Bollettino storico pisano* 51 (1982): 295–309.
80. Ilardi, *Renaissance Vision,* 9.
81. Quoted in ibid., 82.
82. David Whitehouse, *Medieval Glass for Popes, Princes, and Peasants* (Corning, NY: Corning Museum of Glass, 2010), 60.
83. For reference to Dante and Saint Catherine of Siena see Roberta J. M. Olson, "Lost and Partially Found: The Tondo, a Significant Florentine Art Form," *Artibus et Historiae* 14, no. 27 (1993): 32. For reference to untidy edges, see Olszewski, "A Possible Source for the Triptych," 7.
84. Works by Pietro Lorenzetti that feature buildings with circular windows include the following: *Beata Umiltà Transports Bricks to the Monastery,* ca. 1341, Uffizi Gallery, Florence; *Entry of Christ into Jerusalem,* ca. 1320, Lower Church, San Francesco, Assisi; *Predella Panel with The Annunciation to Sobac*, ca. 1328–1328, Pinacoteca Nazionale, Siena; and *The Miracle of the Ice*, ca. 1341, Staatliche Museen, Berlin.
85. While these two examples date to the late fourteenth or early fifteenth century, it is likely that they reflect earlier practices.

86. As Carolly Erickson argues, the early modern world was filled with various layers of reality, where one's visionary imagination was equally valuable to sensory data and when supernatural entities could cross the threshold into the natural world. In other words, the medieval mind did not see the spiritual realm and the earthly one as entirely divorced from one another. There were moments when one seeped into the other and these were vitally important, informative, and monumental events. This multidimensionality of the early modern outlook may prove a fruitful context when analyzing the many different levels of reality presented in the art of the trecento. From the reliquaries with their combination of different media to the frescoes that simultaneously reference the earthy world and the divine realm, the complex qualities of the art may have reflected this dynamic way of viewing the world. See Carolly Erickson, *The Medieval Vision: Essays in History and Perception* (New York: Oxford University Press), passim, esp. 213–19.

References

Alberti, Leon Battista. *On Painting*. Translated by John R. Spencer, rev. ed. New Haven, CT: Yale University Press, 1966.

Alighieri, Dante. *The Divine Comedy*. Edited by David H. Higgins, Translated by Charles H. Sisson. Oxford: Oxford University Press, 2008.

———. *The Convivio*. Edited and Translated by Andrew Frisardi. Cambridge: Cambridge University Press, 2018.

Angenendt, Arnold. "Relics and Their Veneration." In *Treasures of Heaven: Saints Relics, and Devotion in Medieval Europe,* edited by Martina Bagnoli, Holger A. Klein, C. Griffith Mann, and James Robinson, 19–28. New Haven, CT: Yale University Press, 2010.

Antoni, Tito. "Note sull'arte vetraria a Pisa fra il Tre e il Quattrocento." *Bollettino storico pisano* 51 (1982): 295–309.

Avery, C. Louis. "Early Christian Gold Glass." *The Metropolitan Museum of Art Bulletin* 16, no. 8 (August 1921): 170–75.

Bagnoli, Martina. "The Stuff of Heaven: Materials and Craftsmanship in Medieval Reliquaries." In *Treasures of Heaven: Saints, Relics, and Devotion in Medieval Europe*, edited by Martina Bagnoli, Holger A. Klein, C. Griffith Mann, and James Robinson, 137–47. New Haven, CT: Yale University Press, 2010.

Bagnoli, Martina, Holger A. Klein, C. Griffith Mann, and James Robinson, eds. *Treasures of Heaven: Saints, Relics, and Devotion in Medieval Europe*. New Haven, CT: Yale University Press, 2010.

Bajetto, Fiorenza. "Un trentennio (1941–1973) di studi sul Cantico di Frate Sole. Bibliografia ragionata." *L'Italia francescana* 49 (1974): 5–62.

Belting, Hans. *Likeness and Presence: A History of the Image Before the Era of Art*. Chicago, IL: University of Chicago Press, 1994.

Beretta, Marco. *The Alchemy of Glass: Counterfeit, Imitation, and Transmutation in Ancient Glassmaking*. Sagamore Beach, MA: Science History Publications, 2009.

Borsook, Eve. *Medieval Mosaics: Light, Color, Materials*. Milan: Silvana, 2000.

Brilliant, Virginia. "A Framework for Devotion in Trecento Siena: A Reliquary Frame in the Cleveland Museum of Art." *Peregrinations: Journal of Medieval Art and Architecture* 4, no. 3 (Spring 2014): 66–94

Burke, Robert Belle. *The Opus Majus of Roger Bacon*. Vol. 2. Whitefish, MT: Kessinger Publishing, 2002.

Bynum, Caroline Walker. *The Resurrection of the Body in Western Christianity 200–1336*. New York: Columbia University Press, 1995.

Carboni, Stefano. *Venice and the Islamic World: 828–1797*. New York: Metropolitan Museum of Art, 2007.

Crosby, Alfred W. *The Measure of Reality: Quantification and Western Society, 1250–1600*. Cambridge: Cambridge University Press, 1997.

Cullen, Christopher M. "Alexander of Hales." In *A Companion to Philosophy in the Middle Ages*, edited by Jorge E. Gracia and Timothy B. Noone, 104–8. Oxford: Blackwell Publishing, 2005.

Dickson, Gary. "The Crowd at the Feet of Pope Boniface VIII: Pilgrimage, Crusade and the first Roman Jubilee (1300)." *Journal of Medieval History* 25, no. 4 (1999): 279–307.

Edgerton, Samuel Y. *The Renaissance Rediscovery of Linear Perspective*. New York: Basic Books, 1975.

———. *The Heritage of Giotto's Geometry: Art and Science on the Eve of the Scientific Revolution*. Ithaca, NY: Cornell University Press, 1991.

———. *The Mirror, the Window, and the Telescope: How Linear Perspective Changed Our Vision of the Universe*. Ithaca, NY: Cornell University Press, 2009.

Erickson, Carolly. *The Medieval Vision: Essays in History and Perception*. New York: Oxford University Press, 1976.

Fisher, John. "Hugh of St. Cher and the Development of Mediaeval Theology." *Speculum* 31, no. 1 (Jan., 1956): 57–69.

Frugoni, Chiara. *Medioevo sul naso: Occhiali, bottoni e altre invenzioni medievali*. Rome: Laterza, 2001.

Gerevini, Stefania. "*Sicut crystallus quando set objecta soli*: Rock Crystal, Transparency, and the Franciscan Order." *Mitteilungen des Kunsthistorischen Institutes in Florenz*, 56, no. 3 (2014): 255–83.

———. "*Christus Crystallus*: Rock Crystal, Theology, and Materiality in the Medieval West." In *Matter of Faith: An Interdisciplinary Study of Relics and Relic Veneration in the Medieval Period*, edited by James Robinson and Lloyd de Beer with Anna Harnden, 92–99. London: British Museum Press, 2014.

Gibbs, Robert. *Tomaso da Modena: Painting in Emilia and the March of Treviso, 1340–80*. Cambridge: Cambridge University Press, 1989.

Hahn, Cynthia J. *Strange Beauty: Issues in the Making and Meaning of Reliquaries*. University Park, PA: They Pennsylvania State University Press, 2012.

Hills, Paul. *The Light of Early Italian Painting*. New Haven, CT: Yale University Press, 1987.

Huizinga, John. *The Waning of the Middle Ages*. Garden City, NY: Doubleday Anchor, 1954.

Ilardi, Vincent. *Renaissance Vision from Spectacles to Telescopes*. Philadelphia: American Philosophical Society, 2007.

Johnson, Timothy J. "Francis and Creation." In *The Cambridge Companion to Francis of Assisi*, edited by Michael J. P. Robson, 143–60. Cambridge: Cambridge University Press, 2012.

Kanter, Laurence B. "A Rediscovered Panel by Fra Angelico." *Paragone* 599 (January 2000): 3–13.

Kennedy, Vincent Lorne. "The Moment of Consecration and the Elevation of the Host." *Mediaeval Studies* 6 (1944): 121–50.

King, Catherine. "The Arts of Carving and Casting." In *Siena, Florence, and Padua: Art, Society and Religion: 1280–1400*, edited by Diana Norman, vol. 1, 97–121. New Haven, CT: Yale University Press, 1995.

Klein, Holger. "Eastern Objects and Western Desires: Relics and Reliquaries between Byzantium and the West." *Dumbarton Oaks Papers* 58 (2004): 283–314.

Lindberg, David C. "Lines of Influence in Thirteenth-Century Optics: Bacon, Witelo, and Pecham." *Speculum* 46, no. 1 (Jan. 1971): 66–83.

———. *Theories of Vision from Al-Kindi to Kepler*. Chicago: University of Chicago Press, 1976.

Leclercq, Eloi. *The Canticle of Creatures: Symbols of Union*. Translated by Matthew J. O'Connell. Chicago: Franciscan Herald Press, 1978.

Lesnick, Daniel R. *Preaching in Medieval Florence: The Social World of Franciscan and Dominican Spirituality*. Athens, GA: Georgia University Press, 1989.

Lubbock, Jules. *Storytelling in Christian Art from Giotto to Donatello*. New Haven, CT: Yale University Press, 2006.

Macfarlane, Alan, and Gerry Martin. *Glass: A World History*. Chicago: University of Chicago Press, 2002.

Maginnis, Hayden. "Assisi Revisited: Notes on Recent Observations." *The Burlington Magazine* 117, no. 869 (August 1975): 511–17.

———. *The World of the Early Sienese Painter*. University Park, PA: Pennsylvania State University Press, 2001.

Mann, C. Griffith. "Naddo Ceccarelli, Reliquary Tabernacle with Virgin and Child." In *The Walters Art Gallery: Guide to the Collection*, 24. London: Scala Books, 1997.

———. "Relics, Reliquaries, and the Limitations of Trecento Painting: Naddo Ceccarelli's Reliquary Tabernacle in the Walters Art Museum." *Word and Image* 22, no. 3 (July–Sept. 2006): 251–59.

———. "Reliquary Triptych with the Annunciation, Saint Ansanus, the Adoration of the Magi, and the Crucifixion." In *Treasures of Heaven: Saints, Relics, and Devotion in Medieval Europe*, edited by Martina Bagnoli, Holger A. Klein, C. Griffith Mann, and James Robinson, 204–5. New Haven, CT: Yale University Press, 2010.

Newhauser, Richard G. "Peter of Limoges, Optics, and the Science of the Senses." *Sense and Society* 5, no. 1 (2010): 28–44.

Norman, Diana. "Those Who Pay, Those Who Pray and Those Who Paint: Two Funerary Chapels." In *Siena, Florence, and Padua: Art, Society and Religion: 1280–1400*, edited by Diana Norman, vol. 2, 169–94. New Haven, CT: Yale University Press, 1995.

Olson, Roberta J. M. "Lost and Partially Found: The Tondo, a Significant Florentine Art Form." *Artibus et Historiae* 14, no. 27 (1993): 31–65.

Olszewski, Edward. "A Possible Source for the Triptych, Lunette, and Tondo Formats in Renaissance Paintings." *Notes in the History of Art* 28, no. 2 (Winter 2009): 5–11.

Recht, Roland. *Believing and Seeing: The Art of Gothic Cathedrals.* Chicago: University of Chicago Press, 2008.

Rowlands, Eliot W. "Sienese Painted Reliquaries of the Trecento: Their Format and Meaning." *Konsthistorisk Tidskrift* 48 (1979): 122–38.

Rubin, Miri. *Corpus Christi: The Eucharist in Late Medieval Culture.* Cambridge: Cambridge University Press, 1991.

Strehlke, Carl Brandon and Machtelt Brüggen Israëls, eds. *The Bernard and Mary Berenson Collection of European Paintings at I Tatti.* Florence: Villa I Tatti in collaboration with Officina Libraria, 2015.

Tachau, Katherine. *Vision and Certitude in the Age of Ockham: Optics, Epistemology and the Foundations of Semantics, 1250–1345.* Leiden: E. J. Brill, 1988.

Thomas of Celano. *The First Life of Saint Francis* in *The Lives of Saint Francis of Assisi.* Translated by A. G. Ferrers Howell. London: Methuen & Co., 1908.

White, John. *Art and Architecture in Italy: 1250–1400.* New Haven, CT: Yale University Press, 1993.

Whitehouse, David. *Medieval Glass for Popes, Princes, and Peasants.* Corning, NY: Corning Museum of Glass, 2010.

6 Verre Églomisé Reliquaries

Pietro Teutonico and Tommaso da Modena

In many ways this chapter is a continuation of the discussion about gilded glass from Chapter 3 and, accordingly, a summary of that chapter's main points is useful. As previously mentioned there was a fourteenth-century revival of a form of ancient sandwich glass with imagery etched into a gilded panel called *verre* églomisé. And as examples from the late-thirteenth through early-fifteenth century demonstrate, gilded glass was frequently used to decorate sites which marked a convergence between the heavenly and earthly spheres. Bernardo Daddi's painting of the Madonna enshrined in Orcagna's tabernacle, for instance, functioned as a conduit between God's divine will and his faithful flock by seemingly responding to prayers with miraculous deeds. In the case of the tomb of Saint Dominic, the *Arca*, Nicola Pisano used gilded glass to commemorate the saint's body, his most precious holy relic. Like all primary relics, that is, relics consisting of saints' body parts, Dominic's corpse embodied profound connections to both the spiritual and the mundane. These relics were human remains and therefore firmly part of the material world. At the same time, they had profound spiritual significance as the human being in question had been canonized and thus considered to be of divine stature.

The present chapter addresses similar themes. Its case studies are all reliquaries that feature verre églomisé and the imagery inscribed on the glass shares similar formal qualities with the gilded panels by Giotto's workshop (Figure 3.6), Paolo di Giovanni Fei (Figure 3.7), and Lorenzo Monaco. The artworks that follow contain highly modeled figural scenes etched into the gilding and present the viewer with a unique meditative opportunity because the pictorial light and surface light intermingle within a single artwork.

The reason this chapter has been separated from the earlier discussion is because these reliquaries also feature something not found in previous examples: a unique combination of gilded glass and transparent glass. In fact, in most cases, the portions of verre églomisé are positioned directly adjacent to sections of transparent glass. Therefore, it was important to first establish an understanding of the perception and history of transparent glass, as done in Chapters 4 and 5. The reliquaries' juxtaposition of transparent glass and gilded glass results in a complex interplay of formal qualities and symbolic meaning.

Verre Églomisé Reliquaries

The Metropolitan Museum's *Reliquary Diptych* (Figure 6.1) is a representative example of a large group of mass-produced verre églomisé reliquaries from the fourteenth century made in central Italy for the Franciscan order.[1] It is a diptych, meaning it has two wooden panels joined by hinges along the inner edges. Each panel consists of a wooden frame and glass panel. When open, it is approximately eight inches wide and seven inches tall. The backs of the panels, visible when closed, each feature a red cross outlined in gold outline dividing the panel into four roughly equal quadrants. It was likely used for personal devotion by a member of the Franciscan order or a wealthy pilgrim who may have obtained it as a souvenir from Assisi or another important Franciscan site.[2]

The Metropolitan's diptych uses the verre églomisé technique to illustrate the Nativity and Crucifixion in the center of each panel and the Angel Gabriel and Virgin Annunciate in lunettes atop each panel. The borders framing the main scenes are also filled with verre églomisé imagery. The Nativity is surrounded by symbols of the four evangelists and two unidentified saints, while portraits of Saints Peter, Paul, Francis, and Clare, along with two other unidentified saints, frame the Crucifixion. Interspersed between these saints' portraits are a series of rectangular fields without gilded imagery.[3] These small, rectangular sections are left untreated and thus function like small windows revealing a variety of relics, including those of Mary Magdalene and the True Cross, along with the relics' authentics, or identifying labels.

The extant reliquaries demonstrate that verre églomisé was used to decorate a variety of portable reliquary types, including diptychs, triptychs, roundels, and

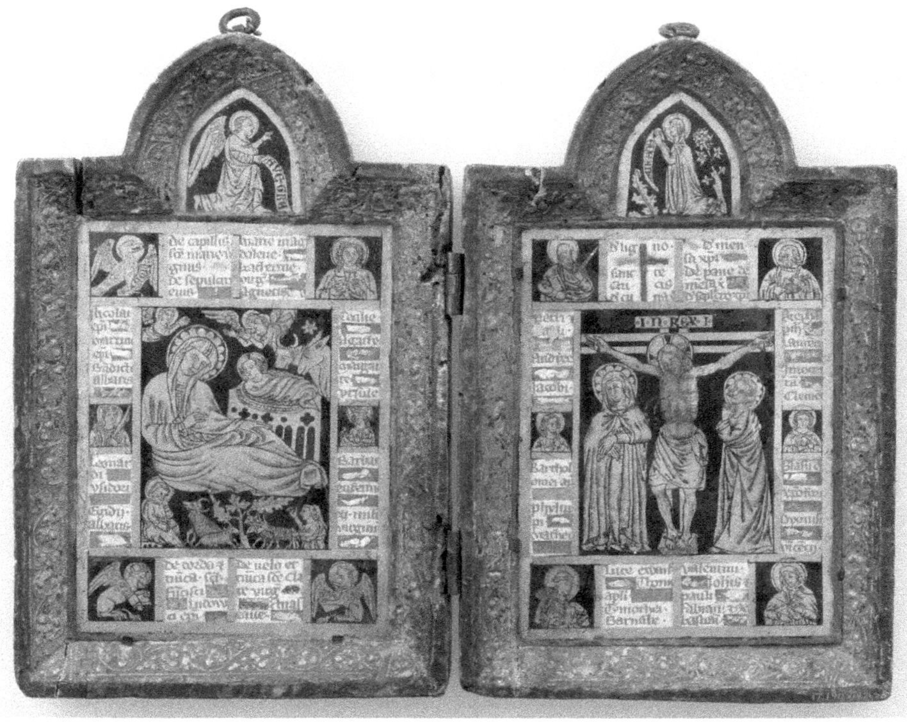

Figure 6.1: Italian Workshop, *Reliquary Diptych,* late-14th century, The Metropolitan Museum of Art, New York, Gift of J. Pierpont Morgan, 1917 (17.190.982). Source: Public Domain, The Metropolitan Museum of Art (CC0).

processional crosses or panels.[4] Most of these reliquaries were products of fourteenth-century Italian workshops in Umbria or the Marches with some exceptions dating to the late twelfth century and early fifteenth century.[5]

As mentioned in Chapter 3, Cennino Cennini describes verre églomisé as a process wherein an artist applied gold leaf to portions of a glass panel, inscribed imagery into the gold leaf in reverse, and then protected the work with a black or colored backing.[6] To understand the role an artist might have played, it is useful to compare Cennini's instructions for this technique to other passages in his treatise. In his discussion of stained-glass windows, for example, Cennini includes advice on designing and painting the glass but omits instructions for making stained glass. This omission suggests that Cennini's intended audience, the artist, was not expected to cast the glass, only design and paint it. Similarly, when giving directions for verre églomisé, Cennini does not provide details on how to cast the glass panel. He does, however, include descriptions for every other step of the process, suggesting that perhaps his reader was to perform all these steps.

Also contained within Cennini's discussion is a brief comment on the reception and application of the technique. He notes that verre églomisé is "indescribably attractive, fine, and unusual, and this is a branch of great piety, for the embellishment of holy reliquaries." Missing from Cennini's texts is further elaboration on on why verre églomisé was considered so "indescribably attractive, fine and unusual," why was it associated with such "great piety," and, perhaps most pertinent to the present discussion, why it was associated with reliquaries. But one can make some inferences.

One could posit that the "attractive" and "fine" qualities to which Cennini refers were related to the shimmering, golden highlights of the gilded imagery. With its almost supernatural sparkle, the verre églomisé Crucifixion scene appears at once part mosaic, part metalwork, and part miniature, certainly an "unusual" trait for the devotional art of this time. The "great piety" associated with verre églomisé is equally difficult to explicate with extant documentation but it may find partial explanation in the precedents discussed in Chapter 3 and in the technique's ancient applications. As mentioned in Chapter 3, early Christians in Rome used roundels decorated with gilded glass to decorate utilitarian objects like bowls and plates and then repurposed them into grave markers. In the case of early Christian martyrs, these panels of gilded glass would have functioned as secondary relics, meaning, they were given special status because they had close contact with a saint.

This chapter contributes an important component to the ongoing dialog about the reception of verre églomisé and its association with reliquaries as referenced by Cennini. At the same time, because the case studies presented here span the entire fourteenth century, this discussion also joins the debate about the impact of the Black Death on artistic practice. Like the artworks by Paolo di Giovanni Fei and Lorenzo Monaco, the reliquaries in this chapter question Meiss's claim because they were made after the plague but, yet, display continued interest in naturalistic imagery. Furthermore, at least two of the works were made in Siena, a city particularly impacted by the devastation of the Black Death.[7] Finally, this chapter may shed even more light on the optical interests and artistic patronage of the Franciscans.

The Franciscan Connection

The frequent appearance of Saint Francis of Assisi and other saints related to the Franciscan order in the verre églomisé reliquaries suggests that many of these objects were commissioned by that religious order. The Umbrian origin of these works also supports this theory, as this region was the home of the Order of the Friars Minor. More evidence of their Franciscan patronage derives from the fact that many reliquaries display strong compositional similarities to the frescoes in San Francesco of Assisi. Like the Assisi frescoes, the imagery depicted in the verre églomisé panels features the new, naturalistic style so often associated with the artistic trends of the fourteenth century. Figures display three-dimensional modeling and there are often suggestions of a background setting and vernacular details. Finally, many objects are still to be found in Franciscan collections today.

The materials used to make the reliquaries—primarily wood and glass—certainly would have been appreciated by most of the fourteenth-century Franciscans. Unlike other religious orders at the time the friars minor were committed to a vow of poverty.[8] Michael J. P. Robson argues that, because of Francis's lavish upbringing, he was well acquainted with the temptations and problems associated with wealth and found it imperative to integrate a vow of poverty into his order.[9] According to Francis, for one to truly follow in Christ's footsteps, poverty was essential. In his *Later Rule*, the *Regula bullata*, which served as the basis for the Franciscan lifestyle, Francis emphasized the value of poverty, noting that the brothers may not receive any money, either directly or indirectly:

> As payment for their work they may receive whatever is necessary for their own bodily needs and [those of] their brothers, but not money in any form; and they should do this humbly as is fitting for servants of God and followers of most holy poverty.

For Francis it was clear; poverty was important because it brought one closer to the "self-emptying" state of Christ, a state comprised of a combination of humility and love.[10] However, for Francis's later followers, the matter was not as straightforward.

The Franciscan commitment to poverty evoked great debate and dissension throughout the thirteenth and fourteenth centuries.[11] Many of Francis's followers, who would come to be known as the Spirituals, or *fraticelli*, remained deeply committed to his vow of poverty.[12] The majority of the order's leaders, sometimes referred to as "the community," believed, however, that the order needed to adapt to its growing role within medieval society by accumulating wealth in order to finance a more established presence in the major cities of Europe and sustain greater pastoral care of the people.[13] Because of their limited use of gold leaf and other relatively inexpensive materials, the verre églomisé reliquaries could have appropriately honored the sacred relics through modest means and thus avoided evoking the controversial issue.

The physical properties of glass may have also had symbolic meaning that made this a suitable medium for decorating reliquaries within a Franciscan context. In the second version of his *Letter to the Faithful*, Francis describes the Incarnation of Christ as the moment when the Son of God "received the flesh of humanity and our frailty."[14] Thus, the fragile glass panel, so easily broken and destroyed, could have served as a reminder of the transience of this human life for one well versed in Franciscan theology. Furthermore, the method of making glass—that is, the transformative process of turning lowly earthly materials into a substance worthy of Heavenly Jerusalem and almost on par with rock crystal—could have resonated with the dual nature of Christ as both fully human and fully divine.[15]

The theme of Christ's duality was well suited to the object's function as a reliquary since the relics represented both the spiritual and physical nature of the commemorated saint. Perhaps not surprisingly this theme is also reflected in the majority of narrative scenes found in the reliquaries. Many of the reliquaries feature the Annunciation, Nativity, and Crucifixion—that is, episodes in Christ's life that involved moments of transition between holy and human states of being, when Christ was conceived, born, and died. The sight of the Crucifixion would have held even greater significance for Francis's followers because Francis of Assisi received the stigmata, bodily wounds that matched those incurred by Christ while nailed to the cross.

Finally as discussed throughout this book, of all the early modern religious orders, the Franciscans would have been some of the most receptive to the optical associations and implications of glass. Several members of the order or individuals associated with the order—most famously Saint Bonaventure, Alexander of Hales, Roger Bacon, John Pecham, John Duns Scotus, and William Ockham—made important contributions to the science of optics and many others referenced optical theory in their theological writings.[16] The order's deep affinity for how study of the natural sciences informs one's faith is captured in the words of Bonaventure, who wrote, "Behold how the Divine Wisdom lies hidden in sense perception, and how wonderful is the contemplation of the five spiritual senses in the light of their conformity to the senses of the body."[17]

The Franciscan emphasis on the visual sense is reflected in some of the order's most famous works of literature and visual art. As mentioned in the last chapter, the *Meditations on the Life of Christ* sought to engage their reader's visual imagination and foster devotion through its descriptive langague. Further evidence of the Franciscan privileging of visual media is found in the extensive artistic program at San Francesco in Assisi which not only used the naturalistic style to mimic the natural world but also incorporated many references to optical devices throughout its walls.

As described in Chapter 5, optical instruments and glass can be found throughout the decorations in the Lower Church of San Francesco in Assisi as well. One can recall, for instance, how Giotto painted Prudence holding a mirror to convey the power of her insight in the central vault and Simone Martini integrated his stained glass into his frescoes in the Chapel of Saint Martin.[18] In a later example, Andrea de' Bartoli da Bologna painted a philosopher holding a pair of eyeglasses and another holding a magnifying glass in his scene of *The Philosophers Confronting Saint Catherine* in the burial chapel of Cardinal Albornoz. And finally, convex glass backed with silver meant to resemble stars in the night sky, cover the ceiling of the Lower Church (Figure 2.5).[19]

Along with their interest in the sense of sight Franciscans studied the nature and theology of its optical counterpart, light. As mentioned in Chapter 3 Saint Bonaventure wrote extensively about the theoretical implications of light in his commentary on the *Sentences of Peter Lombard*, identifying the main components of light as lux, lumen, and color.[20] As in the case of the earlier examples of verre églomisé, comparing Bonaventure's description to the formal characteristics of the Metropolitan Museum's *Reliquary Diptych* reveals interesting connections between the different aspects of light and the visual experience provided by the reliquaries.

As the essential nature and source of light, lux was comparable to God in a spiritual analogy and a candle flame or the sun in a more practical corollary.[21] Lumen, the emanation of lux, was most commonly observed as the aura of light radiating out from a candle flame or as the rays of the sun. In terms of a symbolic interpretation, lumen could have been associated with Christ because he was seen as the ultimate emanation from God and the vehicle by which God's light entered the world. These analogies are corroborated by Bonaventure's description of Christ's relationship to God in his *Tree of Life*,

> In this eternal kingdom, all good and perfect gifts come down in plenty and abundance from the Father of Lights (James 1:17) through Jesus Christ, who is the superessential Ray … For he is a pure effusion of the brightness of the power of the omnipotent God.[22]

For Bonaventure, color was the manifestation of light resting on an opaque surface after making its way through a transparent medium such as air or clear glass. Color, therefore, was the entity that made things visible to the eye. As with lux and lumen, color also finds a visual parallel in the *Reliquary Diptych* and similar verre églomisé reliquaries. First, there is actual pigment found throughout the scenes; for instance, red pigment is applied to the back of the panel depicting the blood of Christ. But, color is also found within the text of the authentics and, though not necessarily colorful, in the relics. Thus, the candle's lux emanated through the air and glass in the form of lumen and, as color, it made visible the relics' shape, texture, and other physical qualities. Within the specific context of the *Reliquary Diptych*, color enabled one to see evidence of this phenomenon in its highest form through the presence of the relics.[23]

While many informative connections between theology and light can be found within the reliquaries Bonaventure's system lacks a visual counterpart to the golden sparkling reflections created by the gilded glass. As previously discussed in Chapter 3, Bartolomeo da Bologna seemed to respond to this issue by adding another type of light to Bonaventure's system, namely, splendor.[24] Unlike color, which fell on the transparent glass and enabled one to study the physicality of the relics, splendor negated the physical characteristics of the panel. When light reflected off the gilded glass panel, reflections obscured one's view of the glass's surface and the gold leaf. In the form of splendor, then, light took on an almost supernatural quality making it a particularly suitable symbol for God's divinity.[25] Therefore, while color may have represented the divine as experienced through the natural, observable world, splendor could have reflected the presence of God in more supernatural, unobservable forms such as in the case of miracles or visions.

There are other reasons that the highlights may have been of particular interest to both the makers and viewers of these reliquaries. In a manner similar to the gilded glass artworks in Chapter 3, the reliquaries could have also served as visual manifestations of the moment of visual contact by evoking the notion of Plato's fiery rays. For, although the intromission model of vision was becoming more dominant over the course of the fourteenth century, the notion of the fiery rays remained a valuable concept in certain circles. What follows considers how these reliquaries—with their complex combination of various types of glass—may have been uniquely equipped to simultaneously resonate with both the intromission and extramission models of visual perception.

Windows to Relics and a Mirror for the Divine

The different treatments of glass—verre églomisé and transparent—within reliquaries such as the Metropolitan Museum of Art's diptych may initially seem to be at aesthetic odds with one another. On the one hand, the portions of transparent glass allowed ambient light to penetrate the glass and illuminate the small relic fragments and authentics. On the other hand, the areas of the glass decorated with gilding would have produced glowing golden reflections evocative of God's presence when struck by light. By combining these two different treatments the reliquaries simultaneously served as a window to the relics and as a mirror of the divine, making them well suited to contemporary visuality and, by extension, very powerful devotion aids.

As described in Chapter 5, a reliquary with relic windows made of transparent glass could have compelled the viewer to look *through* the glass's surface to the relic fragments of wood, cloth, and hair. Because of the small size of the relics, the trecento viewer needed to get close and perhaps even strain one's eyesight in order to see the relics. Such intense gazing likely lent the process of viewing the panel a very intimate and intense tone. An emphasis on careful, engaged, and active looking would have been further emphasized because, in order to obtain validation of the relics, the viewer needed to be literate and read the text written on the authentics. Further stress on the visual experience derived from the fact that, unlike earlier forms of relic worship these relics did not offer the viewer an opportunity to touch, kiss, or ingest the holy fragments. Set behind the glass plate, these precious fragments were *only* available to one's sense of sight.

In a general way this type of spiritual devotion—so dependent on one's physical sense of sight—may be seen as evocative of an empirical viewing experience and perhaps even a harbinger of the growing interest in scientific investigation usually equated with a Renaissance outlook.[26] Encountering the relics in this context allowed the viewers to use their eyes to see, or to know, the human side of Christ and the saints by engaging with evidence of their material form as a means of contemplating the divine. Fourteenth-century viewers could have established a powerful connection with their spiritual role models, and ultimately Christ, through examining material evidence of their shared human experience.

In a manner very different from the transparent portions of glass, the areas of the reliquary's glass decorated with verre églomisé depict canonical Christian narratives and emphasize the surface of the glass panel. When viewing the golden reflections bouncing off the gilded imagery one would have witnessed the sharp, bright instances of light. Paul Hills suggests that such an experience could have been interpreted as a powerful connection between the viewer and the holy image because light and spiritual enlightenment were commonly associated with one another at this time.[27] Such gilded imagery could have reminded the fourteenth-century viewer of more traditional medieval reliquaries which typically concealed relics behind an array of precious materials such as gold in order to evoke the precious nature of the divine.[28] Using their inner mind's eye and spiritual insight the faithful viewer could reflect on the divine nature of God in a more abstract way.

Thus the transparent glass provided viewers with an opportunity to use their sensory perception to examine physical evidence of the saints' human nature while the golden highlights had the potential to abstractly symbolize the divine. By treating the glass in these two distinct ways the artist offered the viewer two dramatically different—but equally valid—visual experiences capable of inspiring spiritual insight: one through reflection on Christ's humanity and the other on his divinity. One wonders if the trecento viewer struggled to reconcile them or, perhaps, privileged one over the other. If the principles behind early modern Christian doctrine can serve as a model, it is likely that a fourteenth-century viewer did not expect cohesion and may have even appreciated a paradoxical approach when contemplating spiritual matters. After all, the reliquary's primary function was to commemorate the relics, the physical evidence of the human nature of Christ and his saints. At the same time the reliquary needed to appropriately celebrate the corresponding celestial spirit. By fusing the two different decorative treatments within a single sheet of glass the reliquary could have elegantly evoked the central mystery of the Christian faith: the nature of Christ as both fully human and fully divine. And, in so doing, it referenced the medieval tradition's penchant for a more abstract approach to the divine with the golden highlights as well as a more empirical approach indicative of burgeoning Renaissance values.

It is possible there were other correlations between the different treatments of the glass and the two main visual theories of the day—intromission and extramission—as well as with the early modern anatomical understanding of the eye. As noted in Chapter 3, the Platonic model of visual theory known as extramission permeated Western European thought from the fourth century through the end of the thirteenth century. Plato theorized that the eye sent out visual beams that coalesced with external light and reached out to the object, resulting in visual perception.[29] Thus when praying before a reliquary, trecento viewers sympathetic to Plato's visual theory would have been able to conceive of the dramatic golden reflections as more than just indications of the holiness of the reliquary. These sharp rays of light bouncing off the glass could have been understood as visual manifestations of Plato's fiery rays emitted from the eye, hitting their object of perception, and returning to the viewer with the input. In this context, the reliquary's gilding offered visual verification of the moment of contact between the viewer and the reliquary thereby ensuring a solid foundation for one's devotional practice. That Plato's visual theory resonated with the golden highlights from the verre églomisé portions of glass rather than the transparent glass is fitting because Plato's philosophical outlook was more concerned with abstract concepts rather than observable phenomenon.

If, however, instead of God's divine nature one aimed to reflect on the humanity of Christ and the saints, the viewer could have reflected on the emerging philosophy and visual theory of Aristotle.[30] As described in Chapter 5, Aristotle's visual theory mostly superseded that of Plato by the beginning of the fourteenth century after its reintroduction to the Latin West through Muslim and ancient Greek optical treatises.[31] Aristotle combined certain aspects of Plato's extramission with its main theoretical opponent, intromission, which maintained that the perceived object sent forth images of itself, known as *eidola,* into the eye. In order to reconcile these two seemingly opposed systems, Aristotle developed what David Lindberg has termed the "mediumistic" visual theory.[32]

Aristotelian theory asserted that the perceived object sent forth *eidola*, or images of itself, into the intervening medium—usually air—and it was through this medium that the visual properties were absorbed and processed by the eye. Lindberg describes Aristotle's position as one not focused on directional rays from the object or eye but rather one that paid more attention "toward the medium between the observer and the visible object."[33] As Lindberg describes it, Aristotle "perceives the absolute necessity of this medium … [which is] diaphanous or transparent, a nature or power found in all bodies, but especially air, water, and certain solid substances."[34] For Aristotle's theory, then, which was ascendant at the time of the fabrication of the Metropolitan's *Reliquary Diptych* and similar verre églomisé reliquaries, a great deal of emphasis and importance was placed on the space between the viewer's eye and the observed object. For visual contact to occur this medium needed to be transparent.

The ability of the Metropolitan's *Reliquary Diptych* to resonate simultaneously with both the Aristotelian and Platonic modes of vision is not a conceptual or devotional weakness but rather constitutes a fitting reflection of a historical context in which both theoretical models were available and possible. Though there was a general shift from the Platonic to the Aristotelian model, the debate between these two models continued from the time of the medieval scholastics until Johannes Kepler. Thus, throughout the fourteenth century, Plato's theory was regarded as valid enough to dispute and therefore a premise worthy of serious consideration.[35]

When contextualizing these visual theories within the sphere of devotional art and the verre églomisé reliquaries it is important to note that visual theories were not purely scientific pursuits but, rather, a means to a spiritual end. Devotional effectiveness superseded optical debate and purely scientific goals. The private nature of prayer afforded the viewer an opportunity to employ whichever theoretical model or models served the ultimate goal of bringing one closer to God. Even if the viewer had exposure to optical theories—which would have at least been somewhat likely if the viewer was an educated member of the Franciscan order—the devotee using the reliquary would not have necessarily been concerned with the specifics of the theoretical models while deep in prayer. Rather, when faced with the potential for spiritual enlightenment the trecento devotee would have likely employed any and all methods at his or her disposal to affect spiritual enlightenment.[36]

As mentioned in Chapter 4 there was another aspect of optical science that could have resonated with the viewer's experience of the transparent glass relic windows, namely, the anatomy of the eye. The ancient scholar who codified much of the early modern understanding of the eye's anatomy was Galen, however, the Latin West obtained most of its information on the eye's structure from Hunain's *Ten Treatises on the Eye*.[37] Hunain described the lens of the eye as an ice-like or crystal-like humor which is uncolored, transparent, luminous, and round with a flattened face.[38] Its transparent nature, he noted, allowed it to quickly receive the color of perceived objects while its round shape prevented breaking or chipping and its flatness provided an ideal surface for receiving visual stimuli.[39] As the seat of vision the crystalline humor occupied the central position in the eye so that it could be served by the other elements.[40]

After the crystal-like humor, the next most important supportive membrane was the one located directly behind the lens. This element was described as the vitreous humor, that is, an component having a glass-like nature. Although the glass-like substance within the eye was not considered the location where vision occurred it was nevertheless crucial for fostering the visual process. The vitreous membrane served the crystalline humor by mediating between the seat of vision and the blood vessels of the retina. As Hunain explained, "A substance is most quickly transmitted into the thing which resembles its own nature most closely."[41] The eye's glasslike substance was ideally suited to serve the crystalline seat of vision because the material of glass was similar to, but not identical to, the properties of crystal.

Knowledge of the eye's glasslike component could have resonated with the glass panels of the reliquaries. As the faithful gazed at the relics through the glass they could imagine that the glass panel fostered a form of visual contact. One thinker supporting the notion that vision was a form of contact was Alexander of

Hales. In the thirteenth-century debate over the benefits of viewing the Elevation of the Host, Hales argued that visual contact was sufficient to receive the host's virtues. He explained that because the Eucharist bestowed spiritual, immaterial benefits, the sense of sight was the most appropriate sense organ with which to receive them, as it was the most immaterial of the five senses.[42] The belief that viewing the relics constituted contact with them would have been especially helpful to one viewing the small fragments within the verre églomisé reliquaries.

While it is difficult to ascertain the extent of critical engagement among trecento viewers with the aforementioned optical theories, the Franciscans were some of the most interested in studying and capitalizing on an understanding of optics for their devotional pursuits. As noted elsewhere in this book, the Franciscan order, the likely patrons and viewers of many of the verre églomisé reliquaries, had a strong penchant for visual media and optics.[43] Therefore questions about the relationship between their reliquaries and their notion of vision are important ones.

The Beatific Vision and Viewing God Through a Glass Darkly

One of the most compelling doctrines to consider in light of the Franciscan Order's optical interests is that of the Beatific Vision. The *visio Dei* was the moment after death when one came into contact with the divine. As famously described by Saint Paul in 1 Corinthians 13:12, "For now we see through a glass, darkly; but then face to face: now I know in part; but then shall I know even as also I am known."[44] Though the message may seem straightforward—while living an earthly existence one's knowledge of God is imperfect and perfect knowledge is only possible in heaven—around the 1330s church leaders, especially Franciscans, intensely debated when this experience of mystical union occurred.[45] Some friars followed the position of Pope John XXII and believed that it happened after the Last Judgment, meaning that there was a waiting period between one's individual death and the Last Judgment. The opposing view, the position promoted by Pope Benedict XII's *Benedictus Deus,* mandated that if individuals were properly cleansed of their sins in purgatory they could experience the *visio Dei* directly after their individual death.[46]

Further complicating an understanding of Paul's comment is the fact that the translation has been interpreted in various ways. Edward Peter Nolan's essay in *Now Through a Glass Darkly: Specular Images of Being and Knowing from Virgil to Chaucer* works to clarify the debate.[47] Nolan describes possible translations as "through a glass darkly," "in a mirror dimly," and "puzzling reflections in a mirror" and argues that, if translated literally, it is likely more correct that Paul's passage refers to a mirror rather than a sheet of glass. However Nolan also points out that many translators have opted to use the word glass instead for its lyrical beauty and because it may in fact bring the reader closer to Paul's original meaning. Basing his analysis on the Latin translation—as this was the source of the medieval commentaries—Nolan deconstructs the passage with particular attention to its prepositions. He concludes that there is no precise or absolute translation possible and thus the passage holds several interpretations simultaneously. Rather than seeing this as a mistake or problem of some sort, Nolan argues that this is very fitting for a passage which tries to explain a mystery when he writes "that the utterance written to clarify an enigma is itself an enigma."[48]

Medieval and Renaissance artists employed various strategies when depicting this moment. Two cases of greatest interest to the present study incorporate depictions of glass panels and were made by artists, namely Giotto and Ambrogio Lorenzetti, who may have been particularly engaged within the network of glass. In around 1340 Ambrogio depicted a mirror in his cycle of the *Allegory of Bad Government* in the council chamber of the Council of the Nine in Siena's Palazzo Pubblico as mentioned in Chapter 5.[49] In the Sala della Pace fresco Lorenzetti painted the three primary vices of Vanity, Pride, and Avarice hovering around the enthroned personification of Tyranny. The personification of Vanity is dressed in sumptuous garb and holds a branch in one hand and a small, golden, circular mirror in the other.[50] Like the classical figure of Narcissus she tilts her head and gazes at her captivating reflection in the small mirror. Her elaborate golden headdress is an adornment worthy of Venus herself. Vanity's virtuous counterpart on the adjacent wall is Hope who, instead of being absorbed with superficial earthly appearances, views God's face and his golden virtuous halo directly. Ambrogio's depiction, therefore, is divided into two episodes, one on the wall representing the notion of good government and one on the adjacent wall depicting the concept of bad government making it clear that this pairing represents fallible versus perfected vision.

Giotto employs a different strategy in his depiction of the Beatific Vision in the pinnacle of the Baroncelli altarpiece, now separated from the main panel and housed in the San Diego Museum of Art (Figure 2.7). The upper cusp of the altarpiece features God the Father at the center and two groups of three angels gazing up towards him the lower corners. In each grouping two angels hold their hands up to the eyes, as if to shield a bright light, while the third angel holds what appears to be a circular piece of glass and looks through this device. It is not entirely clear what this object is meant to represent. Compared to other fourteenth-century depictions of optical technologies treated in more detail in Chapter 5, this object is larger than the typical magnifying glass and mirror. It may represent an enlarged mirror for the sake of legibility as in the case of Ambrogio Lorenzetti's Massa Marittima altarpiece discussed shortly. Another possibility is that this was Giotto's attempt to interpret Paul's ambiguous passage and not a representation of an available technology. Whatever Giotto's motivation it is important to note that he incorporated a glass object, suggesting that, at least in some cases the passage was clearly not interpreted as referring to a metal mirror.

It is also important to recall that Giotto had depicted an accurate rendition of a convex mirror already by 1305. In the Arena Chapel, Giotto painted a series of grisaille personifications of virtues and vices running along the bottom of the two sidewalls. The figure of Prudence sits behind a broad desk and holds a measuring compass in one hand and a small, circular mirror in the other.[51] She wears simple attire. Free from the distraction of an elaborate wardrobe she directs her attention to the mirror and the open book propped up on a pedestal before her. She has two faces; the one in front is youthful while the one in the back appears older. Her attributes identify her ability to see the past, present, and future with equal clarity because, as described by Cicero and well known in the medieval period,

> Prudence is the knowledge of what is good, what is bad and what is neither good nor bad. Its parts are memory, intelligence, [and] foresight. Memory is the faculty by which the mind recalls what has happened. Intelligence is the faculty by which it ascertains what is. Foresight is the faculty by which it is seen that something is going to occur before it occurs.[52]

Thus the older face symbolized knowledge of the past, the measuring compass and book represented one's understanding of the present, and the mirror signified knowledge of the future, or foresight.

Giotto or his workshop painted another mirror with similar significance about twenty years later in the vault of the Lower Church of Assisi.[53] Within the fresco depicting the Franciscan virtue of obedience in the vault near the crossing one finds Prudence holding a mirror out to a friar and another possible mirror set in front on a table and seen only from the back. As with the figure of Prudence in the Arena Chapel, this figure's identity is revealed through an accompanying inscription as "S.PRUDENTIA," indicating this is a representation of Sacra Prudentia or Holy Prudence, an alias for Wisdom. However, while their names are very similar, during the ancient and medieval periods there was a distinct difference between Prudence as practical knowledge and Sacred Prudence, which had a higher, usually more spiritual nature. In the apocryphal writing from the Septuagint now referred to as the Book of Wisdom or the Wisdom of Solomon, Sacred Prudence was described as "the brightness of eternal light, and the unspotted mirror of God's majesty, and the image of his goodness."[54] The device of the mirror was again associated with forms of intellectual or spiritual insight, knowledge, and understanding.

The notion that the mirror could reveal divine truth was already in place during early Christian times. In the third century, Origen wrote,

> Wisdom is also called the stainless mirror of the … working of God. … For as the image formed in a mirror unerringly reflects all the acts and movements of him who gazes on it, so would Wisdom have herself to be understood when she is called the stainless mirror of the power and working of the Father.[55]

Thus, the symbolic mirror like the actual mirror offered more visual information than was previously available.

In the 1330s Ambrogio Lorenzetti depicted a mirror in his monumental public altarpiece of the *Maestà* for Massa Marittima with meaning similar to the mirrors painted by Giotto. Ambrogio's mirror also provided a view of something not typically seen or possible to view without some type of divine aid. As Herbert Kessler observes, Ambrogio's mirror "captures the image of what is opposite while it shields the viewer from direct exposure."[56] Ambrogio painted an enthroned Madonna and Child positioned atop three stone steps. Upon each step sits a personification of a theological virtue. From bottom to top they are Faith, Hope, and Charity. Faith holds and gazes into a mirror, which originally had silver leaf covering the now visible red bole.[57] During the 1970s this painting was closely examined and it was discovered that the reflection in the mirror originally featured two faces back-to-back and a soaring bird. Faint traces of the bird and faces are still visible in incised lines in the bole. The bird in the reflection is likely a dove, the symbol of the Holy Spirit, suggesting that Faith's mirror revealed the mystery of the Holy Trinity.[58]

During the Middle Ages the stainless mirror became associated with the Virgin Mary.[59] Because Mary miraculously conceived Christ through the power of the Holy Spirit, she was able to retain her virgin state despite giving birth. Mary was likened to a mirror that allowed light to pass through the glass and reflect off the lead backing without breaking.[60] The earliest glass mirrors' ability to reflect the viewer and their surroundings must have provided a strikingly unusual visual experience for the early-modern viewer since the mirror could extend one's visual field in a new direction and provide more visual information than was possible with natural eyesight alone, even if the early modern mirror was not of the quality of modern ones.

The rich symbolic potential of the mirror was likewise felt in the literary world. The word *speculum* was increasingly being used in titles and texts of medieval literature and making frequent appearances in one of the most famous texts from the early modern era, the *Divine Comedy.*[61] Dante's work from around 1310 describes a pilgrim's journey for the revelation of divine light or, put another way, a divine vision.[62] Dante the Pilgrim traverses the dark lands of Hell and Purgatory before his arrival in Heaven where he, for a fleeting brief moment, glimpses the divine glory of God. Dante refers to a mirror no less than thirty times throughout the poem.[63] One such instance occurs when the pilgrim encounters Rachel and Leah in Canto 26 of Purgatory. The former, emulating Venus, uses the mirror for vain purposes and the latter, in the manner of Faith, uses it contemplate deeper truths.[64] Then in Canto 2 of Paradise Beatrice, having assumed the role of Dante's guide, explains that the dark spots on the moon are not due to different matter reflecting light differently but, rather, to the object's ability to reflect divine illumination solely to the extent that God allowed it. She makes her divine logic clear to Dante by performing an experiment involving three mirrors.[65] The frequent reference to mirrors in the poem is not a coincidence. Like many of the visual artists discussed in this book Dante was also engaging with optical theory and incorporating it into his work. As mentioned in Chapter 5, Dante wrote his own description of the visual process in *Convivio* III, ix., referencing an intromission model of vision and noting the importance of the transparent intervening medium and even equating this medium to glass.

It is useful to consider verre églomisé reliquaries such as the Metropolitan's diptych within the context of Saint Paul's comment on the Beatific Vision and the presence and reception of mirrors and glass because it was thought that physical vision reflected its spiritual counterpart. Fittingly for such a multivalent devotional tool, the artist of the *Reliquary Diptych* used a single sheet of transparent glass as both a window to relics and a mirror of the divine, in order to inspire faith and transcendental insights into the promise of the Beatific Vision and direct knowledge of God. Similar lines of analysis can apply when examining Tommaso da Modena's reliquary however, because of its even more elaborate display context, this panel of glass can provide even more potential insight.

Tommaso da Modena's Reliquary and Ugo da Panciera's *Treatise on Perfection*

Another example of a reliquary using the verre églomisé technique that seems to draw on an optical understanding for reaching spiritual ends is the panel executed largely by Tommaso da Modena and now in the Walters Art Museum, Baltimore (Figure 6.2) which was originally part of a diptych or triptych. The extant 18-inch-high wooden panel from around 1360 features a variety of media and styles. Naturalistically painted portraits of saints and symbols of the evangelists along with insets of marble and ceramic line the panel's outermost boarder.[66] The tabernacle is surmounted by a piece of verre églomisé at the pinnacle featuring the Virgin Annunciate in gold leaf with black backing.[67] At the center of the work is another panel of verre églomisé depicting the Crucifixion also backed with black pigment. This scene is surrounded by a series of saints inscribed into gold leaf and backed with red or green pigment.[68] Interspersed between the saints are portions of the glass panel that have been left clear. These areas of transparent glass reveal a group of small relic fragments and their authentics displayed behind the glass panel.[69] While Tommaso is credited with the paintings and the overall artistic direction, the gilded glass panels likely belonged to the family of gilded glass responsible for the aforementioned examples and thus arrived pre-made. Therefore what follows here will not focus on analysis of the verre églomisé panels but rather on how Tommaso's paintings resonate with them and with concurrent optical technology, Christian theology, and Franciscan devotional practice.

Tommaso's panel features characteristics that appear to respond to a visually inclined trecento culture by harnessing optical theory to enhance spiritual devotion. This may not be surprising given that Tommaso painted several other artworks which reference optics and suggest he was acutely aware of many different types of optical aides. As mentioned in Chapter 5, about a decade prior to making his reliquary panel, Tommaso painted a series of famous Dominicans in the monastery at San Nicolò in Treviso with optical devices. These include a portrait of Cardinal Hugh of St. Cher with eyeglasses, Cardinal Nicholas of Rouen with a magnifying glass, and Pietro Isnardo da Chiampo of Vicenza with a reading mirror.[70] For the same patrons Tommaso painted a portrait of Saint Jerome in the adjacent church with yet another reading mirror. Depicting church fathers wearing or using optical tools suggests that these inventions were not viewed as inherently manipulative or deceptive but, rather, that their helpful properties could be used in service of spiritually virtuous ends. After all, the primary objective of Tommaso's reliquary was not to appeal to the scientific community but a spiritual

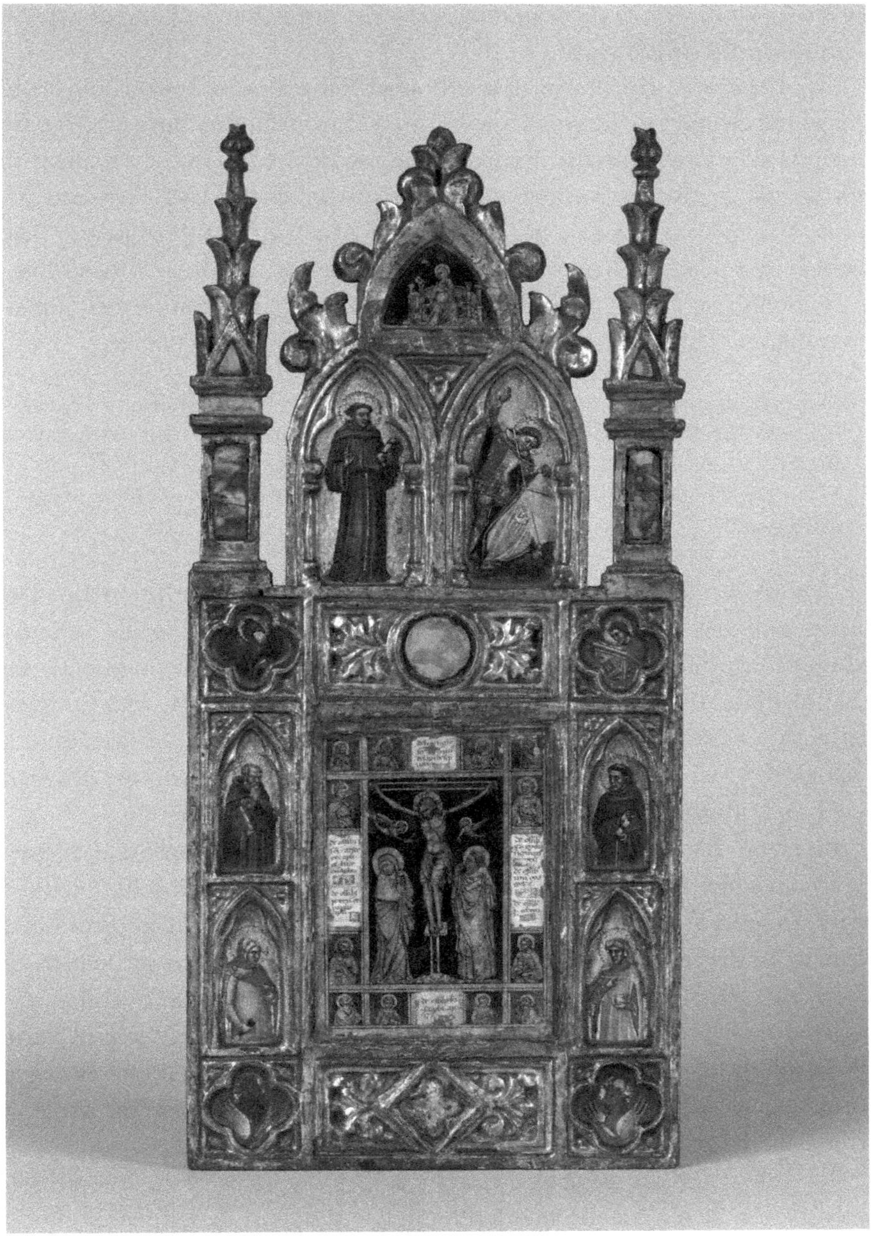

Figure 6.2: Tommaso da Modena, *Wing of a Reliquary Diptych with the Crucifixion and Saints*, ca. 1355–70, The Walters Art Museum, Baltimore, Acquired by Henry Walters with the Massarenti Collection, 1902 (37.1686). Source: Public Domain, The Walters Art Museum (CC0).

viewer—most likely a Franciscan judging by the prominent full-length depiction of Francis in the upper right.

The Franciscan context is especially important because only a few decades prior to the creation of Tommaso's reliquary a Franciscan friar described the steps one must take to elevate their understanding to divine insight in a manner that finds several significant correlations within this reliquary. Ugo Panciera's *Trattato della perfezione (Treatise on Perfection)*, written in in approximately 1320, describes the process of contemplating the divine as similar to the crafting of an artistic image; as the concepts became clearer, they took on formal qualities, becoming more modeled, colored, and fleshed out. As Panciera notes,

> In the first moment when the mind begins to think about Christ, Christ seems written in the mind and in the imagination; in the second, Christ seems to have been sketched; in the third, he seems to have been under-drawn and under-painted; in the fourth, he seems to have been colored and his flesh to have been painted; in the fifth he seems incarnate and *rilevato*.[71]

In a manner similar to Panciera's developing mental image Tommaso's panel features different media and imagery with various degrees of detail, coloration, and modeling. Not only that but the variety of media seems to consciously offer visual manifestations of the various stages of developing insight with an inverse relationship between the amount of visual definition and the complexity of the theological doctrine. In other words, the more complex the idea, the less clarity there is in both the conceptual understanding and visual description.

Panciera's final and fifth stage, where the idea is "incarnate and rilevato," equates a vivid visual image to conceptual clarity. When looking for a corollary within Tommaso's panel one finds the most accessible and straightforward elements to be the relics. The holy fragments are not depictions of something else; rather, they *are* sacred entities. The relics are literally "incarnate and rilevato" as described in Panciera's fifth stage and, although they ultimately referred to the saint residing in heaven, they did so in very familiar terms by presenting the viewer with pieces of objects such as a tooth, hair, cloth, or wood. The authentics further verified and documented the identity of these fragments.

The full-length portraits of the saints and the symbols of the evangelists in the outermost border of the wooden frame are the most developed illusionistic imagery in the piece. Painted with a relatively wide range of colors, high degree of spatial perspective, and naturalistic details these figures are some of the largest in the panel. In addition, these painted portraits have carefully modeled physiognomies, individualized facial expressions, and varied gestures. They wear different hairstyles and are shown in a range of formats (e.g., full-length, 3/4-length, and half-length). With their detailed description these figures could have easily correlated with Panciera's fourth level, that is, the level where the image "seems to have been colored and his flesh to have been painted."

Moving inward to the border of verre églomisé saints one finds that these portraits are also highly individualized but not modeled to the same degree as their painted counterparts on the wooden frame.[72] These verre églomisé saints are smaller, their poses are more limited, and the use of color is restricted. The figures are comprised of black lines and gold leaf while the red and green color is restricted to the background. Therefore the verre églomisé saints could have resonated with Panciera's third stage, the under-painting and under-drawing, where the imagery is less developed but still has a degree of modeling and color.

Moving inward still to the very center of the gold glass, Christ on the cross is modeled in a manner similar to the adjacent saints but the scene's background is completely black. The only color used is the small amount of red describing the blood of Christ. This portion of the panel contains the least color and correspondingly this scene depicts one of the most complex Christian beliefs—the death and Resurrection of Christ—allowing it to serve as a visual manifestation of the second, sketched phase of Panciera's model. The analogy to a sketch becomes even more appropriate when one observes that the individualized lines etched into the gold leaf are clearly visible, giving the image of Christ a very linear quality.

During the initial phase of Panciera's developmental scheme, one does not yet have a clear enough understanding to envision the concept. As Panciera explains, at this moment the difficult concept is still in "written" form. It has not yet taken mental shape. The authentics are certainly one aspect of the panel that could relate to this description, however, these written labels are intimately connected to the relics. For this reason it is useful to expand consideration of the "written" stage beyond the confines of the devotional panel to the Bible, the written word of God and the basis for all Christian belief. As it is said in the introduction of John's gospel, "In the beginning was the Word, and the Word was with God, and the Word was God." With this in mind, it is helpful to consider the most elusive story or concept in the Christian faith, namely the Beatific Vision, as a fitting correlation as above in the analysis of the Metropolitan Museum's *Reliquary Diptych*.

By combining the painted figures with the imagery in the verre églomisé panel, Tommaso da Modena's reliquary tabernacle gave visual form to complex conceptual frameworks such as that underlying Panciera's treatise. The tabernacle harnessed both the advantages of panel painting as well as those of verre églomisé to offer multiple contemplative avenues. Whatever developmental stage of

Panciera's scheme the viewer was at, there was a corresponding visual element within Tommaso's panel on which to set one's gaze. Through the various lighting effects and the different treatments of the glass, the reliquary panel could have also accommodated whether one wanted to contemplate the human qualities of Christ or the divine aspects of God. By combining these seemingly opposing qualities Tommaso's panel resonated with the unique qualities of trecento visuality and devotion and, in doing so, demonstrates for the modern reader the intersections among fourteenth-century art, religion, and science intersected.

Conclusions on Verre Églomisé Reliquaries: Reflections of God

Glass, in its various forms, provided the trecento artist a multifunctional artistic medium that could inspire complex devotional strategies. While the transparent glass used for relic windows could have resonated with the growing interest in the sense of sight and the value of empirical observation, the gilded imagery provided an opportunity to contemplate the more metaphysical aspects of one's faith through the golden imagery of the verre églomisé. Artists seem to have realized that when it came to the notion of the Beatific Vision not even the finest transparent glass or the strongest magnifying lens was helpful and they therefore employed a different visual system entirely. And yet, at the same time, there appears to have been a strong association between the medium of glass and the visual process and such connections may have inspired the artists' use of this material in their devotional artworks, thereby fostered greater contemplation of it and it's symbolic meaning.

Notes

1. For a list of other reliquaries of a similar type see Dillian Gordon, "The Mass Production of Franciscan Piety: Another Look at Some Umbrian *Verres Églomisès,*" *Apollo* 140, no. 392 (1994): 33–42.
2. In his analysis of a similar object Hugh Hudson suggests such works may have also served as wedding gifts, see Hugh Hudson, "From Assisi to Melbourne: Friar Pietro Teutonico's Nativity; Crucifixion Reliquary Diptych in the National Gallery of Victoria," in *Interpreting Francis and Claire of Assisi: From the Middle Ages to the Present*, eds. Constant J. Mews and Claire Renkin (Melbourne: Broughton Publishing, 2010), 242–54.

3. The unidentified saint in the center of the left border is a male figure holding the martyr's palm and the one in the center of the right panel is female, potentially Saint Veronica holding the sudarium.
4. According to Georg Swarzenski, "The Localization of Medieval Verre Églomisé in the Walters Collection," *Journal of the Walters Art Gallery* 3 (1940): 55 there are 60 extant pieces but it is unclear what types of objects are included in this inventory. Cristina De Benedictis describes a large catalog of Umbrian verre églomisé in Cristina De Benedictis, *Devozione e produzione artistica in umbria* (Florence: Edifir, 2010). No exhaustive catalog Renaissance verre églomisé currently exists. Because the reliquaries were portable and made of fragile glass one could assume there was once many more examples that were destroyed.
5. For more on the production and history of type of work see Cristina De Benedictis, *Devozione e produzione,* passim; Dillian Gordon, "The Mass Production of Franciscan Piety: Another Look at Some Umbrian Verres Églomisés," *Apollo* 140, no. 392 (1994): 33–42; and Irene Hueck, "Ein umbrisches Reliquiar im Kunstgewerbemuseum Schloss Kopenick," *Forschungen und Berichte. Staatliche Museen zu Berlin* (1991): 183–8. For more information on the works dating to the late duecento see Swarzenski, "Medieval Verre Églomisé," n. 1 & 55.
6. Cennino Cennini, *Il libro dell'arte*, trans. Daniel V. Thompson, Jr. (New Haven: Yale University Press, 1933), text on pp. 107–8, translation on pp. 112–14. For Italian text, see also Carlo and Gaetano Milanesi, eds., *Il libro dell'arte o trattato della pittura, di Cennino Cennini da Colle Valdelsa; di nuovo pubblicato con molte correzioni e coll'aggiunta di più capitoli tratti dai codici fiorentini* (Florence: Felice Le Monnier, 1859).
7. Additionally of interest is the fact that Siena was frequently associated with verre églomisé. Scholars have noted but not yet explained why verre églomisé objects were commonly designated as "Sienese" in earlier documents. As Swarzenski notes in his discussion of Francesco di Vanuccio's double-sided processional image in "Medieval Verre Églomisé in the Walters Collection," 50, "No one knows, however, why in old collections … *églomisé* objects were labeled with an obstinate predilection as 'Sienese.'"
8. Michael J. P. Robson, ed., *The Cambridge Companion to Francis of Assisi* (Cambridge: Cambridge University Press, 2012), 3–4.
9. Francis not only understood the potential problems with wealth but also anticipated its strong temptation and, perhaps for this reason, dictated in his Rule that friars should not be allowed to even touch money. Robson, *The Cambridge Companion to Francis of Assisi*, 6.
10. David Burr, *The Spiritual Franciscans: From Protest to Persecution in the Century after Saint Francis* (University Park: Pennsylvania State University Press, 2001), 2.
11. For more on this debate see Burr, *The Spiritual Franciscans*, 261–277 and more recently, David Burr, "Effects of the Spiritual Franciscan Controversy on the

Mendicant Ideal," in *The Origin, Development, and Refinement of Medieval Religious Mendicancies*, ed. Donald S. Prudlo (Leiden: Brill, 2011), 277–305.

12. Illustrating the complicated nature of such an issue is the fact that even those who were committed to the virtue of poverty experienced problems stemming from the order's attitude toward money. The story of John Pecham's rise to archbishop illustrates this situation: though Pecham was deeply committed to his Franciscan vow of poverty, he nevertheless had to spend a great deal of money to travel to Rome to be mitered and then to travel to his post in Canterbury. This debt had long-lasting negative effects for the archbishop according to Burr, *The Spiritual Franciscans*, 7 and Decima Douie, *Archbishop Pecham* (Oxford: Clarendon Press, 1952), 49.
13. For a discussion of the term used to describe the group which opposed the view of the spirituals, see Burr, "Spiritual Franciscan Controversy," 279.
14. Regis J. Armstrong and Ignatius C. Brady, *Francis and Clare: The Complete Works* (New York: Paulist Press, 1982), 66–67.
15. For how glass might have related to the study of matter in the ancient world see Marco Beretta, *The Alchemy of Glass: Counterfeit, Imitation, and Transmutation in Ancient Glassmaking* (Sagamore Beach, MA: Science History Publications, 2009).
16. Roger Bacon and John Pecham both joined the Franciscan order around 1250 in Oxford: David C. Lindberg, "Lines of Influence in Thirteenth-Century Optics: Bacon, Witelo, and Pecham," *Speculum* 46, no. 1 (Jan 1971): 67. For more on John Duns Scotus, see Katherine H. Tachau, *Vision and Certitude in the Age of Ockham: Optics, Epistemology and the Foundations of Semantics, 1250–1345* (Leiden: E. J. Brill, 1988), 55.
17. Emma Therese Healy, ed. and trans., *Reductione artium ad theologiam*, Vol. 1, 2nd ed. (New York: Franciscan Institute, 1955), 49.
18. For more on Giotto's mirror see Herbert L. Kessler, "Speculum," *Speculum* 86, no. 1 (Jan. 2011): 1–41.
19. Hayden Maginnis, "Assisi Revisited: Notes on Recent Observations," *The Burlington Magazine* 117, no. 869 (Aug. 1975): 512.
20. Arthur Zajonc, *Catching the Light: The Entwined History of Light and Mind* (Oxford: Oxford University Press, 1993), 97–98 and Edward Grant, ed., *A Source Book in Medieval Science* (Cambridge, MA: Harvard University Press, 1974), 383.
21. John Pecham, a thirteenth-century Franciscan natural scientist known for his work on optics, equated lux to a candle flame. Hills, *The Light of Early Italian Painting,* 11.
22. Bonaventure, "The Tree of Life," 170–171.
23. For a similar description of Bonaventure's description of color, and the other qualities of light, see Umberto Eco, *Art and Beauty in the Middle Ages* (New Haven, CT: Yale University Press, 1985), 50.
24. Hills, *The Light of Early Italian Painting*, 11. For more on Bartolomeo da Bologna (sometimes written Bartolomeo di Bartoli da Bologna, Bartholomew of Bologna, or Bartholomaei de Bononia) see Graziella Federici-Vescovini, *Le teorie della luce e della*

visione ottica dal IX al XV secolo: Studi sulla prospettiva medievale e altri saggi (Perugia: Morlacchi Editore, 2003), 29–33.

25. A similar interpretation of light reflecting off gold as seen in the medieval tradition can be found in what Paul Hills describes as surface light. Hills, *The Light of Early Italian Painting*, 3–28.
26. This discussion stems from Paul Hills's discussion of the shift from an interest in materiality and what he calls "surface light," to a more naturalistic style and lighting that illuminates the scene, which he defines as "pictorial light." See Hills, *The Light of Early Italian Painting*, 3–28. The visual display of relics was not commonplace in medieval reliquaries. When describing the *Reliquary of the Staff of Saint Peter* Hahn, 287, notes, "Typical of reliquaries made in the western medieval world, this golden staff holds its relic tightly and invisibly, inaccessible to either devout or skeptical eyes." Cynthia Hahn, "What Do Reliquaries Do for Relics?" *Numen* 57, no. 3/4 (2010): 284–316.
27. In his discussion of the gilded panel's highlights in relation to trecento light and optics, Hills, *The Light of Early Italian Painting*, 18 notes,

    > the lustre moving over the surface ... could be experienced as a personal link between the worshipper's eyes and the sacred image ... the light of Christ's brightness could literally shine upon the eyes; and this physical light entering the eyes would have been intuitively understood as metaphor for spiritual illumination.

28. The description of heaven from Revelations 21 describing Heavenly Jerusalem as a city of gold covered in gems and jewels was well known in the trecento. The association between gold and the divine was further strengthened after the influx of Byzantine icons into Italy in the thirteenth century, which inspired early modern Italy's assimilation of the characteristically Byzantine gilded panels and chrysography. Hills, *The Light of Early Italian Painting*, 25. Gold was even enlisted to work miracles; the early modern physician Gentile da Foligno prescribed a beverage with gold emulsion as a remedy for the plague. This instance is recounted in Millard Meiss, *Painting in Florence and Siena After the Black Death: The Arts, Religion and Society in the Mid-Fourteenth Century* (New York: Harper and Row, 1951), 75 n. 5.
29. David C. Lindberg, *Theories of Vision from Al-Kindi to Kepler* (Chicago: University of Chicago Press, 1976), 5:

    > Visual fire emanates from the eye and coalesces with its like, daylight, to form 'a single homogeneous body' stretching from the eye to the visible object: this body is the instrument of the visual power for reaching into the space before the eye.

30. Hills, *The Light of Early Italian Painting*, 12–3.

31. Hayden Maginnis, *The World of the Early Sienese Painter* (University Park, PA: Pennsylvania State University Press, 2001), 2; David Lindberg, *Studies in the History of Medieval Optics* (London: Variorum Reprints, 1983), 349–50.
32. Lindberg, *Theories of Vision*, 6–9.
33. Ibid., 7.
34. Ibid.
35. Ibid., 122–32.
36. For an excellent overview of the wide range of cultural perceptions of sight and a corresponding bibliography see Cynthia Hahn, "Vision," in *A Companion to Medieval Art: Romanesque and Gothic in Northern Europe*, ed. Conrad Rudolph (Oxford: Blackwell Publishing, 2006), 44–64.
37. Hunain's description of the anatomy of the eye was extremely influential for both Islamic and Italian understanding of the nature of the eye. Lindberg, *Theories of Vision*, 34–41.
38. Ibid.
39. Hunain, *The Book of the Ten Treatises on the Eye,* trans. Max Meyerhof (Cairo: Government Press, 1928), 3.
40. Lindberg, *Theories of Vision*, 34.
41. Hunain, *The Book of the Ten Treatises on the Eye*, 6.
42. Roland Recht, *Believing and Seeing: The Art of Gothic Cathedrals,* trans. Mary Whittall (Chicago: University of Chicago Press, 2008), 71.
43. For more on the Franciscan use of images in devotional practice see Anne Derbes, *Picturing the Passion in Late Medieval Italy: Narrative Painting, Franciscan Ideologies, and the Levant* (New York: Cambridge University Press, 1996). For a discussion of how Franciscans relied more heavily on sensory experience and the Dominicans on logic see David L. Jeffrey, "Franciscan Spirituality and the Growth of Vernacular Culture," in his *By Things Seen: Reference and Recognition in Medieval Thought* (Ottawa: University of Ottawa Press, 1979), 150–1.
44. Other translations of the first portion include the Revised Standard Version's "Now we see in a mirror dimly" or the New English Bible's "Now we see puzzling reflections in a mirror." For a discussion and analysis of the various translations, see Edward Peter Nolan, "Introduction: The Cracked Looking Glass," in his *Now Through a Glass Darkly: Specular Images of Being and Knowing from Virgil to Chaucer* (Ann Arbor: University of Michigan Press, 1990): 1–13.
45. Bernard McGinn, "*Visio dei:* Seeing God in Medieval Theology and Mysticism," in *Envisaging Heaven in the Middle Ages,* eds. Carolyn Muessig and Ad Putter (New York: Routledge, 2007), 16 and Decima Douie, "John XXII and the Beatific Vision," *Dominican Studies* 3, no. 2 (1950): 154–74.
46. Caroline Walker Bynum, *The Resurrection of the Body in Western Christianity 200–1336* (New York: Columbia University Press, 1995), 283–91.
47. Nolan, *Now Through a Glass Darkly,* 1–13.

48. Ibid., 2.
49. For more on this fresco cycle, see, Diana Norman, "'Love Justice, You Who Judge the Earth:' The Paintings of the Sala dei Nove in the Palazzo Pubblico, Siena," in her *Siena, Florence, and Padua: Art, Society and Religion: 1280–1400,* vol. 2 (New Haven, CT: Yale University Press, 1995), 145–67.
50. Paula M. Hancock, "Transformations in the Iconography of the Mirror in Medieval Art" (Ph.D. diss., Emory University, 1988), 155 noted that this is the earliest instance of Vanity holding a mirror.
51. For more on the figure of Prudence and the fresco cycle in addition to numerous high-quality images, see Joachim Poeschke, *Italian Frescoes: The Age of Giotto, 1280–1400* (New York: Abbeville Press, 2005). See also Giuseppe Basile, *Giotto: The Frescoes of the Scrovegni Chapel in Padua* (New York: St. Martin's Press, 2002); Bruce Cole, *Giotto: The Scrovegni Chapel, Padua* (New York: George Braziller, 1993); James Stubblebine, *Giotto: The Arena Chapel Frescoes* (New York: Norton, 1969). For more on the mirror specifically, see Douglas P. Lackey, "Giotto in Padua: A New Geography of the Human Soul," *Journal of Ethics* 9, no. 3-4 (2005): 551–72; Eva Frojmovic, "Giotto's Circumspection," *Art Bulletin* 89 (2007): 195–210. For other important discussions see Laura Jacobus, "Giotto's Annunciation in the Arena Chapel, Padua," *The Art Bulletin* 81, no. 1 (March 1999): 104–05; Laura Jacobus, "Giotto's Design of the Arena Chapel," *Apollo* 142, no. 406 (1995): 39–42; and Laura Jacobus, *Giotto and the Arena Chapel: Art, Architecture, and Experience* (Turnhout: Brepols, 2008).
52. Quoted in Hancock, "Iconography of the Mirror," 108–14. For more on early modern references to Cicero see Edgar Wind, *Pagan Mysteries in the Renaissance* (New Haven, CT: Yale University Press, 1958), 260.
53. On the vault frescoes in the Lower Church at Assisi, see Poeschke, *Italian Frescoes,* 109–13 and Kessler, "Speculum," 16.
54. Book of Wisdom, 7:26.
55. Hancock, "Iconography of the Mirror," 109. For original translation, see Alexander Roberts and James Donaldson, eds., *Ante-Nicene Fathers: Translations of the Writings of the Fathers Down to AD 325,* vol. 4 (New York: Scribner's Sons, 1926), 251.
56. Kessler, "Speculum," 32. For analysis of the other elements, see ibid., 21–33.
57. For a detailed study of this figure see Howard Hibbard, "A Representation of Fides by Ambrogio Lorenzetti," *Art Bulletin* 39 (1957): 137–8 and Norman Muller, "Reflections in a Mirror: Ambrogio Lorenzetti's Depiction of the Trinity," *Art Bulletin* 61, no. 1 (1979): 101–2. For more on the original silver facing see specifically ibid., 101.
58. Considering this in comparison to Ambrogio's mirror in the Palazzo Pubblico it becomes apparent that two mirrors painted around the same time by the same artist could hold two completely divergent meanings. Whether the mirror symbolized good or evil depended on context. Other examples of depictions of mirrors with various meanings support this conclusion. They include: Andrea di Bonaiuto's depiction of a mirror held by Wisdom in the roundel at the pinnacle of Saint Thomas Aquinas's

throne in the Spanish Chapel of Santa Maria Novella from ca. 1366; the mirror held by an angel in the pinnacle of Giotto's Baroncelli Altarpiece from ca. 1333 in San Diego; and Buffalmacco's fresco of Hell in the Pisa Camposanto from ca. 1335 featuring a mirror in the hands of one of the damned souls in the bottom right corner. For general notes on the Spanish Chapel see Poeschke, *Italian Frescoes*, 362–5. For comments on fresco in Pisa, see Luciano Bellosi, *Buffalmacco e il Trionfo della Morte* (Milan: 5 Continents Editions, 2003). The most comprehensive discussions on the symbolism of the mirror are found in Kessler, *Speculum* 1–41 and Hancock, "Iconography of the Mirror," 80–174.

59. Hancock, "Iconography of the Mirror," 114–5.
60. The lead backing was thought to symbolize Mary's humility; see ibid., 114.
61. As Heinrich Schwarz noted, "The metaphorical use in literature of the word *speculum* to indicate the complete and precise image of a theme, although extending back into antiquity, becomes particularly frequent following the thirteenth century." Heinrich Schwarz, "The Mirror in Art," *Art Quarterly* 15 (1952): 103. For more on this see Sister Ritamary Bradley, "Backgrounds of the Title *Speculum* in Medieval Literature," *Speculum* 29, no. 1 (1954): 100–15 and Herbert Grabes, *The Mutable Glass: Mirror-Imagery in Titles and Texts of the Middle Ages and English Renaissance* (Cambridge: Cambridge University Press, 1983).
62. Dante, *Divine Comedy*, in *The Portable Dante,* trans. Mark Musa (New York: Penguin, 1995), 3.
63. H. D. Austin, "Dante and Mirrors," *Italica* 21, no. 1 (March 1944): 13.
64. Dante, *Divine Comedy*, 350. Another interesting connection between the Massa Marittima altarpiece and Dante's poem is the fact that Dante described colored stairs leading to paradise that match those depicted by Lorenzetti, and compares one of these steps to a mirror:

> We reached the steps. White marble was the first,
> and polished to the glaze of a looking glass;
> I saw myself reflected as I was.
> The second one was deeper dark than perse,
> of rough and crumbling, fire-corroded stone,
> with cracks across its surface—length and breadth.
> The third one, lying heavy at the top,
> appeared to be of flaming porphyry,
> red as the blood that spurts out from a vein;

(Purgatory 9, verses 94–102; ibid., 244). As far as I am aware, this connection has not yet been made. If Lorenzetti knew of Dante's work the artist could have been inspired by the steps as well as the symbolism of the mirror in this verse and throughout the poem.

65. For further discussion of the three-mirror experiment see Hancock, "Iconography of the Mirror," 170–2.
66. The painted saints include Francis and Michael in full-length view and Anthony Abbot, Anthony of Padua, and Cosmas and Damian, according to Robert Gibbs, *Tomaso da Modena: Painting in Emilia and the March of Treviso, 1340–80* (Cambridge: Cambridge University Press, 1989), 296.
67. The fact that the Virgin Annunciate is shown without the Angel Gabriel is just one of the compelling reasons many scholars have assumed this panel was one side of a diptych.
68. These attributions are according to Gibbs, *Tomaso da Modena,* 296.
69. The panel contains relics of Saints Andrew, Luke, Peter and Paul on the left side, the Holy Cross and the Holy Sepulcher at the top, the 11,000 Virgin Martyrs and one of the Three Kings on the right side, and Saint James on the bottom, according to Martina Bagnoli and others, eds., *Treasures of Heaven: Saints, Relics, and Devotion in Medieval Europe* (New Haven, CT: Yale University Press, 2010), 203. The authentics are transcribed as follows: "de lingio sancte crucis / de lapide sepulcri; de ossibus sancti andree apostoli et luce evangeliste / de ossibus xi. milia virginium / de ossibus unius magorum; de ossibus Jacobi apostoli R." Gibbs, *Tomaso da Modena*, 296.
70. Ilardi, *Renaissance Vision,* 18–20.
71. For the original Italian version, see Arrigo Levasti, *Mistici del Duecento e del Trecento* (Milano: Rizzoli, 1960), 273. For translated version, see Lars R. Jones, "Visio Divina- Donor Figures and Representations of Imagistic Devotion; The Copy of the *Virgin of Bagnolo* in the Museo dell'Opera de Duomo, Florence," in *Italian Panel Painting of the Duecento and Trecento,* ed. Victor M. Schmidt, Studies in the History of Art 61, Center for Advanced Study in the Visual Arts Symposium Papers XXXVIII (CASVA Symposium, Florence and Washington, June and October 1998) (Washington, DC, New Haven, CT, and London: Yale University Press, 2002), 44.

> Nel primo tempo nel quale la mente comincia colle infrascritte circunstanzie di Cristo a pensare, Cristo pare nella mente e nella imaginativa scritto. Nel secondo pare disegnato. Nel terzo pare disegnato e ombrato. Nel quarto pare colorato e incarnato. Nel quinto pare incarnato e rilevato. Tanto ha la mentale virtù attiva di perfezione, quanto può colla corporale virtù attiva regnare. Questo stato della mentale virtù attiva, colla corporale virtuosa azione, merita, per divina giustizia, il dono della meditazione e della contemplazione.

72. Such a difference certainly stems from the fact that there is a difference in media; nevertheless, the end result is images with varied levels of modeling and I argue this is the key for the viewer and intended by the patron and artist.

References

Alighieri, Dante. *Divine Comedy.* Translated by Mark Musa. New York: Penguin, 1995.

Austin, H. D. "Dante and Mirrors." *Italica* 21, no. 1 (March 1944): 13–7.

Bagnoli, Martina, Holger A. Klein, C. Griffith Mann, and James Robinson, eds. *Treasures of Heaven: Saints, Relics, and Devotion in Medieval Europe.* New Haven, CT: Yale University Press, 2010.

Basile, Giuseppe. *Giotto: The Frescoes of the Scrovegni Chapel in Padua.* New York: St. Martin's Press, 2002.

Bellosi, Luciano. *Buffalmacco e il Trionfo della Morte.* Milan: 5 Continents Editions, 2003.

Beretta, Marco. *The Alchemy of Glass: Counterfeit, Imitation, and Transmutation in Ancient Glassmaking.* Sagamore Beach, MA: Science History Publications, 2009.

Bonaventure. "The Tree of Life." In *Bonaventure: The Soul's Journey into God, The Tree of Life, The Life of St. Francis,* ed. Ewert Cousins, 118–68. New York, NY: Paulist Press, 1978.

Bradley, Sister Ritamary. "Backgrounds of the Title *Speculum* in Medieval Literature." *Speculum* 29, no. 1 (1954): 100–15.

Burr, David. *The Spiritual Franciscans: From Protest to Persecution in the Century after Saint Francis.* University Park, PA: Pennsylvania State University Press, 2001.

———. "Effects of the Spiritual Franciscan Controversy on the Mendicant Ideal." In *The Origin, Development, and Refinement of Medieval Religious Mendicancies,* ed. Donald S. Prudlo, 277–306. Leiden: Brill, 2011.

Bynum, Caroline Walker. *The Resurrection of the Body in Western Christianity 200–1336.* New York: Columbia University Press, 1995.

De Benedictis, Cristina. *Devozione e produzione artistica in Umbria.* Florence: Edifir, 2010.

Cennini, Cennino. *Il libro dell'arte.* Translated by Daniel V. Thompson, Jr. New Haven, CT: Yale University Press, 1933.

Cole, Bruce. *Giotto: The Scrovegni Chapel, Padua.* New York: George Braziller, 1993.

Derbes, Anne. *Picturing the Passion in Late Medieval Italy: Narrative Painting, Franciscan Ideologies, and the Levant.* New York: Cambridge University Press, 1996.

Douie, Decima. "John XXII and the Beatific Vision." *Dominican Studies* 3, no. 2 (1950): 154–74.

———. *Archbishop Pecham.* Oxford: Clarendon Press, 1952.

Eco, Umberto. *Art and Beauty in the Middle Ages.* Translated by Hugh Bredin. New Haven, CT: Yale University Press, 2002.

Federici-Vescovini, Graziella. *Le teorie della luce e della visione ottica dal IX al XV secolo: Studi sulla prospettiva medievale e altri saggi.* Perugia: Morlacchi Editore, 2003.

Francis of Assisi. *A Letter to the Entire Order.* In *Francis and Clare: The Complete Works.* The Classics of Western Spirituality. Edited by Regis J. Armstrong & Ignatius C. Brady, 55–61. New York: Paulist, 1982.

———. *The Canticle of Brother Sun.* In *Francis and Clare: The Complete Works.* The Classics of Western Spirituality. Edited by Regis J. Armstrong & Ignatius Brady, 37–9. New York: Paulist, 1982.

Frojmovic, Eva. "Giotto's Circumspection." *Art Bulletin* 89, no. 2 (2007): 195–210.

Gordon, Dillian. "The Mass Production of Franciscan Piety: Another Look at Some Umbrian *Verres Églomisés*." *Apollo* 140, no. 392 (1994): 33–42.

Gibbs, Robert. *Tomaso da Modena: Painting in Emilia and the March of Treviso, 1340–80*. Cambridge: Cambridge University Press, 1989.

Grabes, Herbert. *The Mutable Glass: Mirror-Imagery in Titles and Texts of the Middle Ages and English Renaissance*. Cambridge: Cambridge University Press, 1983.

Grant, Edward, ed. *A Source Book in Medieval Science*. Cambridge, MA: Harvard University Press, 1974.

Hahn, Cynthia J. "Vision." In *A Companion to Medieval Art: Romanesque and Gothic in Northern Europe*, edited by Conrad Rudolph, 44–64. Oxford: Blackwell Publishing, 2006.

———. "What Do Reliquaries Do for Relics?" *Numen* 57, no. 3/4 (2010): 284–316.

Hancock, Paula M. "Transformations in the Iconography of the Mirror in Medieval Art." Ph.D. dissertation, Emory University, 1988.

Healy, Emma Therese, ed. and trans. *Reductione artium ad theologiam*. 2nd ed. New York: Franciscan Institute, 1955.

Hibbard, Howard. "A Representation of Fides by Ambrogio Lorenzetti." *Art Bulletin* 39 (1957): 137–8.

Hills, Paul. *The Light of Early Italian Painting*. New Haven, CT: Yale University Press, 1987.

Hudson, Hugh. "From Assisi to Melbourne: Friar Pietro Teutonico's Nativity; Crucifixion Reliquary Diptych in the National Gallery of Victoria." In *Interpreting Francis and Claire of Assisi: From the Middle Ages to the Present*. Edited by Constant J. Mews and Claire Renkin, 242–54. Melbourne: Broughton Publishing, 2010.

Hueck, Irene. "Ein umbrisches Reliquiar im Kunstgewerbemuseum Schloss Kopenick." *Forschungen und Berichte. Staatliche Museen zu Berlin* (1991): 183–8.

al-'Ibādī, Ḥunain ibn Isḥāq. *The Book of the Ten Treatises on the Eye*. Translated by Max Meyerhof. Cairo: Government Press, 1928.

Ilardi, Vincent. *Renaissance Vision from Spectacles to Telescopes*. Philadelphia: American Philosophical Society, 2007.

Jacobus, Laura. "Giotto's Design of the Arena Chapel," *Apollo* 142, no. 406 (1995): 37–42.

———. "Giotto's Annunciation in the Arena Chapel, Padua," *The Art Bulletin* 81, no. 1 (March 1999): 93–107.

———. *Giotto and the Arena Chapel: Art, Architecture, and Experience*. Turnhout: Brepols, 2008.

Jeffrey, David L. "Franciscan Spirituality and the Growth of Vernacular Culture." In *By Things Seen: Reference and Recognition in Medieval Thought*, edited by David L. Jeffrey, 143–60. Ottawa: University of Ottawa Press, 1979.

Jones, Lars R. "Visio Divina–Donor Figures and Representations of Imagistic Devotion: The Case of Bernardo Daddi's Copy of the *Virgin of Bagnolo* in the Museo dell'Opera del Duomo, Florence." In *Italian Panel Painting of the Duecento and Trecento,* edited by Victor M. Schmidt, Studies in the History of Art 61, 30–55. Washington, DC: Yale University Press, 2002.

Kessler, Herbert L. "Speculum." *Speculum* 86, no. 1 (2011): 1–41.

Lackey, Douglas P. "Giotto in Padua: A New Geography of the Human Soul." *Journal of Ethics* 9, no. 3-4 (2005): 551–72.

Levasti, Arrigo. *Mistici del Duecento e del Trecento.* Milan: Rizzoli, 1960.

Lindberg, David C. "Lines of Influence in Thirteenth-Century Optics: Bacon, Witelo, and Pecham." *Speculum* 46, no. 1 (Jan. 1971): 66–83.

———. *Theories of Vision from Al-Kindi to Kepler.* Chicago: University of Chicago Press, 1976.

———. *Studies in the History of Medieval Optics.* London: Variorum Reprints, 1983.

Maginnis, Hayden. "Assisi Revisited: Notes on Recent Observations." *The Burlington Magazine* 117, no. 869 (August 1975): 511–17.

———. *The World of the Early Sienese Painter.* University Park, PA: Pennsylvania State University Press, 2001.

McGinn, Bernard. "*Visio dei:* Seeing God in Medieval Theology and Mysticism." In *Envisaging Heaven in the Middle Ages,* edited by Carolyn Muessig and Ad Putter, 15–33. New York: Routledge, 2007.

Meiss, Millard. "Light as Form and Symbol in Some Fifteenth-Century Paintings." *Art Bulletin* 27, no. 3 (Sept. 1945): 175–81.

Milanesi, Carlo and Gaetano, eds. *Il libro dell'arte o trattato della pittura, di Cennino Cennini da Colle Valdelsa; di nuovo pubblicato con molte correzioni e coll'aggiunta di più capitoli tratti dai codici fiorentini.* Florence: Felice Le Monnier, 1859.

Muller, Norman. "Reflections in a Mirror: Ambrogio Lorenzetti's Depiction of the Trinity." *Art Bulletin* 61, no. 1 (1979): 101–2.

Nolan, Edward P. *Now Through a Glass Darkly: Specular Images of Being and Knowing from Virgil to Chaucer.* Ann Arbor, MI: University of Michigan Press, 1990.

Norman, Diana. "'Love Justice, You Who Judge the Earth': The Paintings of the Sala dei Nove in the Palazzo Pubblico, Siena." In *Siena, Florence, and Padua: Art, Society, and Religion: 1280–1400,* edited by Diana Norman, vol. 2, 145–67. New Haven, CT: Yale University Press, 1995.

Poeschke, Joachim. *Italian Frescoes: The Age of Giotto, 1280–1400.* New York: Abbeville Press, 2005.

Recht, Roland. *Believing and Seeing: The Art of Gothic Cathedrals.* Chicago: University of Chicago Press, 2008.

Roberts, Alexander, and James Donaldson, eds. *Ante-Nicene Fathers: Translations of the Writings of the Fathers Down to AD 325.* Vol. 4. New York: Scribner's Sons, 1926.

Robson, Michael J. P., ed. *The Cambridge Companion to Francis of Assisi.* Cambridge: Cambridge University Press, 2012.

Schwarz, Heinrich. "The Mirror in Art." *Art Quarterly* 15 (1952): 97–118.

Stubblebine, James. *Giotto: The Arena Chapel Frescoes.* New York: Norton, 1969.

Swarzenski, Georg. "The Localization of Medieval Verre Églomisé in the Walters Collection." *Journal of the Walters Art Gallery* 3 (1940): 54–68.

Tachau, Katherine H. *Vision and Certitude in the Age of Ockham: Optics, Epistemology and the Foundations of Semantics, 1250–1345.* Leiden: E. J. Brill, 1988.

Thomas of Celano. *The First Life of Saint Francis* in *The Lives of St. Francis of Assisi.* Translated by A. G. Ferrers Howell. London: Methuen & Co., 1908.

Wind, Edgar. *Pagan Mysteries in the Renaissance.* New Haven: Yale University Press, 1958.

Zajonc, Arthur. *Catching the Light: The Entwined History of Light and Mind.* Oxford: Oxford University Press, 1993.

7

Conclusion

Giotto, Brunelleschi, Alberti, and the Network of Glass

This book's discussion has thus far been organized according to different types of glass (e.g., stained glass, gilded glass, and transparent glass), which is a useful approach because it reveals important connections among similar artistic treatments of glass. No scholarly accounts of which I am aware discuss the fact that Nicola Pisano's *Pulpit* (Figure 3.5) and Simone Martini's *Maestà* (Figure 1.2) both incorporate small gilded panels of glass even though these works are located in two of the most important sites of Siena, the cathedral and town hall respectively. Nor has any prior investigation sought to contextualize these artworks and their use of glass within the other glass related endeavors in that same city, such as the cathedral's prominent stained glass window designed by Duccio (Figure 2.1), the depictions of transparent glass vessels (Figure 2.3) and gilded Cosmati panels (Figure 2.2) in Duccio's high altarpiece, the many reliquaries with glass relic windows made by Sienese artists (Figures 5.1, 5.2, 5.3, & 5.4), and the mirror depicted in *Allegory of Bad Government in the Palazzo Pubblico*.

A limitation of a methodology based on glass types, however, is a de-emphasis on chronology. It is also difficult to trace the many different ways an individual artist engaged with a variety of types of glass. This concluding chapter, therefore, looks to address this by providing a summary of some chronological themes with focus on specific artists, primarily Giotto, Pietro Lorenzetti, and Taddeo Gaddi. Doing so allows one to start to see, and therefore begin to better understand, the role glass may have played in the evolution of early modern Italian visual culture.

In other words this chapter maps the early modern network of glass and reveals the many interrelated glass objects and ideas shaping trecento visuality. It also suggests some ways this network of glass may have impacted fifteenth-century Italian art and culture with particular interest in the ways glass may have contributed to the notion of the perspectival window.

Through Giotto's Eyes

As Chapters 2 and 3 discuss, medieval artists used glass to make Islamic glass reliquary flasks, glass mosaic tesserae, gilded glass Cosmati panels, enamel decoration, and stained glass. Italian artists not only had access to these artistic traditions but also to ancient glass such as early Christian gold glass roundels, especially in cases of artists who were generally interested in reviving other aspects of ancient art like Nicola Pisano and Arnolfo di Cambio. Perhaps it is not surprising, therefore, that some of the earliest instances of glass in early modern Italian art were made by these artists. Nicola Pisano's *Arca di San Domenico* (Figure 3.4) and *Pulpit* (Figure 3.5), both dated to around the third quarter of the thirteenth century, feature an innovative treatment of gilded glass which finds inspiration in a combination of various ancient and medieval art forms. The impact of the *Arca* and *Pulpit*, two prominent monuments located in highly visible settings, appears to have been felt immediately. Only a few years after working as an assistant to Nicola Pisano, Guglielmo da Pisa continued to explore the artistic potential of etched gilded glass panels, using them extensively in a pulpit for San Giovanni Fuorcivitas in Pistoia made around 1270.[1] Arnolfo di Cambio, another member of Nicola's workshop during both projects, also used gilded glass extensively in his monuments made in the decades directly following his time with Nicola Pisano such as the *Tomb of Cardinal Guillaume de Bray* and the *Ciborium* at San Paolo fuori le Mura to name just two.

Arnolfo is an especially interesting case because he is also responsible for the *Madonna and Child* (Figure 1.2) designed for the façade of the Florentine cathedral around 1296–1302, which features glass insets for the Madonna's eyes. Thus this sculpture not only demonstrates another way this artist engaged with glass but also suggests that artists, and not just optical theorists, might have been aware of the notion that the eye's anatomy is composed—at least in part—of a vitreous medium as discussed in Chapter 4. It is possible that a wider public was also interested in the nature of the eye's anatomy. Only a few years after Arnolfo made his Madonna with the glass eyes Dante referenced the connection between glass and the eye in his *Convivio*. While writing about about the visual process Dante notes how visual stimuli enter into the eye, not as whole objects but as images or forms, which travel "through the diaphanous medium, not actually but mentally, as through transparent glass."[2] The importance of a transparent medium in the visual process was also referenced in Aristotle's theory of vision, which—like many other realms of the scientific inquiry at this time—was highly influential during the trecento.

Dante's *Convivio* is also an important source for establishing the availability of glass mirrors by the time that Giotto was painting his depiction of the small round mirror in the hands of Prudence in the Arena Chapel. After comparing the journey taken by imagery into the eye to transmission through clear glass Dante describes how the visual images culminate in the pupil. For Dante the pupil contains a confined body of water that creates a stopping point for the transmission of the visual form. To help explain this phenomenon Dante referenced a mirror specifically made of glass noting, "And in the water that is in the pupil of the eye, this passage of the visual form is bounded—almost like a mirror, which is glass bounded by lead—so that it can pass no further. …"[3] While it is true that Dante never finished this text and therefore these comments were not circulated by this means, his ideas may have been shared in less formal ways. Furthermore, it is salient that Dante wrote this work in the vernacular because it indices Dante's intended audience, the wider literate populous, would have likely been receptive to such debates or perhaps even familiar with them in some sense already.

An early-fourteenth-century chronicle documents that news regarding optical innovations was indeed popular news to the general public. As discussed in Chapter 5, Fra Giordano's 1306 Lenten sermon in Florence contained the announcement of the invention of eyeglasses.[4] In addition to those who attended and heard the news in person many more could have learned about the invention through the widely disseminated written version of the sermons.[5] It is possible that Giotto himself could have come into contact with Fra Giordano or heard about his news. Giotto was recorded as living in the area around Santa Maria Novella by 1301 and thus in close proximity to the church where Fra Giordano was stationed, putting the artist and preacher very close in terms of both time and space.[6] Regardless of the extent of the interaction between Giotto and Fra Giordano one can be fairly certain that the Florentine public, including Giotto, was interested in news about eyeglasses and, if given the opportunity, this specific artist would have likely been receptive to learning more about optical matters.

Giotto's optical interests seem to permeate his work at the Arena Chapel. It might be said that his revolutionary naturalism in Padua appears to embrace the power of empirical observation in its celebration of forms found within the lived human experience. But there are also more overt references to glass as a material. As discussed earlier, Giotto's depiction of Prudence holds a mirror likely made of glass and he also depicted hanging oil lamps made of glass in the illusionistic niches on either side of the chancel arch (Figure 4.6), which will be discussed in more detail shortly. In addition to these representations of glass Giotto also incorporated actual pieces of glass. He inserted circular silver backed glass roundels into the halo of Christ and may have also included similar glass panels in the vault to represent stars in the manner found in the ceiling of the Lower Church in Assisi.[7] He also designed the windows of the space thereby considering the potential of window glass.[8]

Furthermore as Laura Jacobus discusses, Giotto consciously considered and carefully manipulated the relationship between the actual and pictorial lighting conditions in both his design of the building (and its windows) and in the planning of the frescoes of the Arena Chapel.[9] Jacobus demonstrates that Giotto originally intended to include a stained glass window in the center of the upper portion of the chancel wall.[10] Positioned here, just above the depictions of the Angel Gabriel and the Virgin Annunciate on either side of the chancel archway, the light from the window could have operated in a manner similar to Simone Martini's Saint Martin Chapel stained glass windows (Figure 2.4) or Taddeo Gaddi's windows in the Baroncelli Chapel (Figure 2.6) in terms of its symbolic and formal integration with surrounding paintings. However as Giotto had the opportunity to design the entire building and not just the window's surrounding walls, there could have been even greater synthesis between his window and frescoes had the window been completed. For as Jacobus observes, Giotto's proposed stained glass window would have faced the direction of the rising morning sun.[11] This fact, coupled with her findings that the Annunciation scenes had gilding scored into the plaster and that shutters likely covered the other windows allows one to imagine Giotto's intended effect: upon daybreak the symbolically powerful and colorful light entered the dark chapel through the stained glass and this physically enlivened the chapel as a space. This moment would have initiated the space's sacred activities for the day and symbolically activated the holy event represented in the Annunciation scene, which was the *raison d'être* of the holy space.[12] Thus although it was never built, this window nevertheless still serves as an important example of the many ways Giotto engaged with optics, light, and glass.[13]

In addition to his optically charged work at the Arena Chapel Giotto worked with glass in several other commissions. He inset gilded glass into the halo of the *Crucifix* for Santa Maria Novella; designed the famous, but now mostly lost, mosaic of the *Navicella* for Old Saint Peter's Church using gilded glass tesserae; painted depictions of glass lamps in the Peruzzi Chapel; and painted transparent drinking glasses in a fragmentary panel of the *Last Supper* now in Munich's Alte Pinakothek (inventory no. 643). Still other works featuring glass are attributable to Giotto's workshop and therefore may also reflect the master's keen interest in glass. They include a highly naturalistic rendering of Christ done in a triangular panel of *verre églomisé* now in the Bandini Collection (Figure 3.6); the small, round mirror in the scene of the *Allegory of Obedience* in the vault of the Lower Church, Assisi (ca. 1320); and the pinnacle of Baroncelli altarpiece (Figure 2.7) which features angels holding round panels of glass. This brief overview makes clear that although Giotto is traditionally considered a painter he actively engaged with a variety of glass types physically, through artistic representations, or by way of conceptual design. I argue it is not a coincidence that Giotto had a strong penchant for glass and that he is also nearly synonymous with the early modern naturalistic style. If glass indeed operated as a thinking tool in the ways suggested by Alan Macfarlane and Gerry Martin, it is not difficult to imagine that the variety of visual experiences provided by glass afforded Giotto new insights and that these perspectives would have made their way into his work.[14]

With Giotto's wide range of glass related activities in mind it is worth revisiting the two aforementioned illusionistic niches featuring hanging oil lamps. Two registers below the Annunciation scenes just discussed one finds illusionistic spaces that appear to depict perspectivally rendered galleries or choir spaces with hanging glass oil lamps set before lancet windows (Figure 4.6). Many scholars have debated the theoretical function, symbolic meaning, and style of these spaces.[15] Thomas de Wesselow's detailed investigation of these niches, though primarily concerned with stylistic and chronological issues, nevertheless provides an important summary of the debates surrounding them.[16] However, still missing is an interpretation of these spaces—and their lamps and windows—within the context of the network of glass. This is needed because despite Laura Jacobus's findings that these spaces were not original to the chapel's design Giotto nevertheless painted them with the recognition that their location was prominent in terms of the overall design of the chapel.

It is important to note that while they may be the earliest of examples, these hanging oil lamps are not unique in trecento art. Giotto's own fresco depicting the *Ascension of Saint John* from around 1315 in the Peruzzi Chapel of Santa Croce shows a similar, though less elaborate version. At least two more instances are found in the scenes of the *Vision of the Thrones* and *Verification of the Stigmata* from the life of Saint Francis in the Upper Church at Assisi. Finally, it is also important to consider that, while lamps like this could be made of other materials such as silver or bone, judging by the excavations of medieval monasteries and churches those found in religious contexts were most frequently made of glass.[17] Suggesting Giotto's lamps are meant to represent glass lamps is the religious context of this space, this artist's general interest in glass, and the shape of them, which is closely related to the form of blown glass lamps at this time.[18] If this is indeed the case then these spaces may take on new meaning within the context of the network of glass. But that meaning can only be fully appreciated after an examination of works with similar themes by Pietro Lorenzetti and Taddeo Gaddi.

Illusionistic Architecture and Glass Vessels: Pietro Lorenzetti and Taddeo Gaddi

Conclusions similar to those just described are also revealed when analyzing the presence of glass in the oeuvres of Pietro Lorenzetti and Taddeo Gaddi. In short, these artists—also particularly associated with naturalism and illusionism—actively engaged with glass in a variety of different ways. As discussed throughout this book, Pietro Lorenzetti's interaction with glass involved a variety of techniques. He painted depictions of glass objects, used actual glass in his reliquary with relic windows (Figure 5.3), and would have had visual exposure to many examples of stained glass and gilded glass in Siena as described in Chapter 5.

Most relevant for the present discussion, however, is a consideration of Pietro's time in Assisi because his work there coincides with two factors: exposure to prominent examples of artworks with glass and his first work to incorporate glass. When painting his program in the south transept of the Lower Church Pietro would have had the opportunity to encounter Simone Martini's stained glass windows in the chapel of Saint Martin, the silver backed glass panels meant to represent stars in the night sky set into the ceiling, and the activities of Giotto's workshop occurring in the nearby transept, not to mention the representations of glass hanging lamps in the Upper Church and the mirror in the vault of the Lower Church.[19] It is not hard to imagine that Pietro was inspired by the creativity and new ideas he found there and perhaps the same can be said of his optical interests and interests in glass as a medium. For it is within this rich artistic atmosphere that he painted an illusionistic niche containing two transparent glass liturgical vessels.

Pietro Lorenzetti's illusionistic space (Figure 5.6) takes the form of a rectangular shelf or cupboard positioned above an actual doorway lintel. Because of the condition of this fresco, it is difficult to clearly see the details of the bottles and space in general but there is still some observable sense of perspective. The lower surface of the space is not visible but the upper portion, the ceiling of the space, is visible and is consistent with a vantage point from below. Pietro also conveyed illusionistic depth by painting a receding wall on the right side that appears to meet the back wall and create a corner which falls between the two bottles but closer to the bottle on the right. The glass vessels, one of which appears to hold water and the other wine, feature globular bodies with thin vertical necks and elegant sloping spouts. The bottle containing wine has a handle that is conveniently facing outward, readily accessible to an imagined potential user. The bottle with water may also have a handle facing to the right, but this is difficult to discern due to the condition of the fresco. It is also difficult to describe another object to the left of the bottles but this object's rectilinear shape and the context might suggest that it is a book, though a paten would be more logical.

The trompe l'oeil effect conveyed in the space emulates the illusionistic niches by Giotto in the Arena Chapel in some fundamental ways.[20] First, the space is presented as a naturalistic extension of the viewer's reality, thus, both paintings attempt to dissolve the wall's surface and provide a glimpse into an extension of the immediate physical world. In other words, the painting operates like a window, not dissimilar to the notion of the perspectival window later described by Alberti in the following century, a point which will be considered in more detail shortly. And, in both cases, objects made of glass occupy the illusionistic space. Thus, these spaces might be said to represent something of a microcosm of the larger phenomenon described throughout this book wherein the appearance of glass seems to correlate with art that strives to achieve convincing naturalism. The comingling of glass and naturalism is certainly found in a yet a third case study by Taddeo Gaddi.

Taddeo Gaddi's work in the Baroncelli Chapel has been described throughout this book as demonstrating a keen awareness of optical issues. Considering Taddeo's time in Giotto's workshop this should not be surprising. Briefly recalling the types of glass present at this site is still important, however, because among the stained glass windows and the pinnacle for the altarpiece which originally featured angels holding round panels of glass, there is an illusionistic space containing glass objects not unlike those by Giotto and Pietro. Along the base of the wall to the viewer's left upon entering the chapel one finds another trompe l'oeil niche containing transparent glass liturgical vessels (Figure 2.8).

Taddeo's illusionistic cupboard features two transparent glass vessels of similar shapes to those found in Pietro Lorenzetti's niche. The bottle on the left contains water and the one on the right, wine. They are set on the upper shelf, which divides the cupboard's space into two roughly equal spaces. To the left of the glass bottles is what appears to be a gold pyx and below the bottles on the lower shelf is a round object resembling a silver paten leaning on the right wall of the shelf. Because this scene is located near the base of the chapel's wall, and therefore well below a viewer's eye, Taddeo made the lower surfaces of the space visible but not the ceiling of the cupboard. The right wall of the space is visible while that on the left is not, a vantage point consistent with an actual viewer's position as this painting is positioned on the left wall of the chapel.

Interestingly for the purposes of this study the glass objects seem to have been given a place of prominence while the gold pyx and silver paten are relegated to the peripheries of the space. The glass objects' primacy is enhanced by the fact that the gold pyx is set behind the archway and thus partially obscured from view. The paten looks as if it was almost haphazardly set aside. To the left of the paten is a circular shape the meaning of which is unclear. It looks as if this circular shape is approximately the same size of the paten, perhaps suggesting that the silver dish had been resting there in a horizontal position until recently when the platter was propped against the wall. In contrast the glass objects are centered in the upper portion of the space and framed by the middle archway, completely visible aside from some occlusion occurring because one is set slightly behind the other.

In all three of these revolutionary illusionistic spaces—Giotto's in Padua, Pietro's in Assisi, and Taddeo's in Florence—glass objects are the focal point. It is almost as if the artists were attempting to grapple with, embrace, or otherwise explore the notion of transparency and translucence in terms of their subject matter and style. This is a particularly interesting consideration in light of the technique and medium the artists used to make them because both the fresco technique and tempera paint generally call one's attention to the surface.

In painting a fresco the artist would have likely drawn out a preliminary sketch on the penultimate layer of the wall's plaster. After preparing the final layer of plaster and readying their paints the artist would apply a small portion of plaster—only as much as one could paint in day's time—covering the preparatory drawing and creating the surface for the final painting. The artist would paint on this freshly laid plaster while the surface was still wet so that, as the plaster dried, the paint fused with the wall's surface. The difficultly of navigating this method's constraints—needing to work quickly and efficiently—are rewarded with a durable painting that could stand the test of time. However one must pay careful attention to the wall's surface throughout the process. Proper preparation and application of the plaster was required. Planning the amount of plaster to apply—and thus the size of the area to paint—necessitated accounting for conditions such as the amount of daylight and level of humidity in the air. The artist needed to be sensitive to the wall's surface throughout the process. If an artist wanted to revise a previously painted passage, this required chipping away a portion of the wall and beginning the process again. The type of paint used also focused one's attention on the painting's surface. Tempera paint, made from mixing pigments with egg yolk, was opaque and therefore did not lend itself to depictions of translucency. And yet despite these qualities of the paint and the fresco technique's emphasis on surface—or perhaps because of it—these artists seemed interested in creating illusionistic spaces where the painting functioned like a window. In doing so they created spaces where one could contemplate the notion of transparency (in other words, the passage of light) using glass objects as a vehicle or thinking tool within a context that generally inspired thoughts about opacity (or, the stoppage of light).

Trecento Glass, Brunelleschi's Mirror, and Alberti's Window

Thus in the early chronological period covered in this book, roughly from 1260 to 1320, one finds the continued influence of medieval traditions and artworks featuring gilded glass and stained glass as well as an emergence of new glass trends that reflect the empiricism of Aristotle and the introduction of Arabic science. Optical theory that described the eye as glass and optical devices made using glass (e.g., mirrors and eyeglasses) were starting to impact artists, specifically Giotto. Then, around the second quarter of the century, Pietro Lorenzetti and Taddeo Gaddi pursued similar lines of inquiry to those of Giotto. This is not to say that Giotto and his circle were unique. There were related activities happening outside Giotto's immediate orbit as well, though these examples postdate the works of Giotto, Pietro, and Taddeo. Tommaso da Modena is one such example of an artist from the second half of the century who engaged with depictions of optical devices—several in this artist's case—and panels of actual glass, as discussed in Chapter 6.

This brief summary shows that Italian artists were using glass as a point of reference in both pre-plague and post-plague Italian art. Investigations of the material nature of glass and its powerful relationship to the sense of sight and the visual arts did not cease in the aftermath of the Black Death. Rather, over the course of the fourteenth century artists working with glass made some notably naturalistic artwork and used glass panels in a variety of ways. Thus, experimentation with glass and its related symbolism and associations can be found throughout the century. This is an important aspect to consider because it enables one to more easily consider the possibility that the ways trecento artists engaged with glass informed the visual culture of the early fifteenth century.

As Samuel Y. Edgerton has shown, the notion of the mirror and window were central to the development of the rediscovery of linear perspective.[21] While it is true that mirrors and windows could be made of materials other than glass, as this book has sought to demonstrate, glass was unique among other available materials at the time because of its rich spiritual symbolism and scientific associations and these characteristics made it especially fitting for experimentation with illusionistic space and optical inquiry. This suggests that the artistic community and general populous may have associated the material of glass with the reflective prosperities of a mirror whether Brunelleschi actually used a specifically glass mirror when he painted his first perspectival image of the Baptistery, in the demonstration of how this panel demonstrated the laws of perspective, in both cases, or in neither case.[22] The theoretical codification of the artistic practice demonstrated by Brunelleschi found in Alberti's treatise on linear perspective is also based, in fundamental way, on an object that was typically associated with glass, namely the window. Alberti describes the initial stages of creating a perspectival painting as follows: "First, I trace a large quadrangle … with right angles on the surface to be painted, in which place it certainly functions for me as an open window."[23] Though it is difficult to ascertain Alberti's familiarity with the illusionistic niches in the Arena Chapel, Lower Church, and Barconcelli Chapel, the artistic legacy of Giotto and his followers was not unknown to Alberti, Brunelleschi, and the other fifteenth-century pioneers of linear perspective. It is therefore worth considering the possibility that the rediscovery of linear perspective should also be contextualized within the trecento network of glass.

Conclusion: Mapping the Trecento Network of Glass

The resurgence of the glass industry in thirteenth-century Europe produced new technologies such as glass windows, mirrors, and eyeglasses, all of which influenced the way people saw the world on a very practical level. But there were also conceptual implications. Glass lenses and mirrors were used in scientific experiments and the medium served as an especially elegant metaphor for religious beliefs related to divine visions and spiritual insights. Amidst this complex network of glass's scientific and religious associations, visual artists started incorporating glass (i.e., gilded glass, stained glass, and transparent glass) in particularly innovative ways—either as actual panels of glass or as depictions of glass objects—that seem to consciously reference the relationship between glass and the sense of sight. As this book suggests, when one considers these glass related ideas, artworks, and associations as various components participating within a single network of early modern glass, new readings of trecento art are possible, additional questions arise, and future debate is fostered.

Notes

1. See Chapter 3 for more information on the works of Arnolfo di Cambio and Guglielmo da Pisa with gilded glass.
2. Dante Alighieri, *The Convivio*, ed. and trans. Andrew Frisardi (Cambridge: Cambridge University Press, 2018), 173. While it is true that Dante never finished his *Convivio* and the circulation of this specific comment must be questioned, it is salient that Dante wrote this work in the vernacular. This suggests that Dante's intended audience, the wider literate populous, would have likely been receptive to such debates and possibly even already familiar with them.
3. Dante, *The Convivio,* 173.
4. Vincent Ilardi, *Renaissance Vision from Spectacles to Telescopes* (Philadelphia: American Philosophical Society, 2007), 5.
5. Ibid., 21–22. For more on this, see Daniel R. Lesnick, *Preaching in Medieval Florence: The Social World of Franciscan and Dominican Spirituality* (Athens, GA: University of Georgia Press, 1989), 111.
6. Ilardi, *Renaissance Vision*, 22 records that Giordano was in Florence from 1302–1305, again from 1306–1307, and again in 1309. According to John White, *Art and Architecture in Italy: 1250–1400* (New Haven, CT: Yale University Press, 1993), 309, Giotto was listed as living in the parish of Santa Maria Novella in 1301. Though Giotto would soon leave to paint the Arena Chapel in Padua, there may have been contact between the two given their close time in Florence and their proximity to Santa Maria Novella. Although this connection is speculative at this point, it is worth noting that Giotto painted the mirror in the *Arena Chapel* shortly after leaving Florence.
7. Laura Jacobus describes "gilded stars" in the ceiling and in the halo of Christ. For more on this see "Giotto's Annunciation in the Arena Chapel, Padua," *The Art Bulletin* 81, no. 1 (March 1999): 104–5. That these pieces in the halo are made of glass is supported by the reproduction of them and the description of them as "mirrors" in Claudio Bellinato, *Giotto: The Scrovegni Chapel* (Ponzano Veneto: Grafiche Vianello Srl., 2006), 80–81.

8. For more on the windows in the Arena Chapel and the use of crown glass or "bulls-eye" glass see Chapter 5.
9. Laura Jacobus, "Giotto's Annunciation," 104–5. Further supporting is the observation made by Hans Michael Thomas and Giuliano Romano which finds that the sunlight coming through the window falls onto the area of the wall where Giotto depicted the patron dedicating a model of the chapel to the Virgin. See Giuliano Romano and Hans Michael Thomas, "Sul significato di alcuni fenomeni solari che si manifestano nella cappella di Giotto a Padova," *Ateneo Veneto* 178 (1991): 213–56 and Hans Michael Thomas, "Sonneneffekte in der Giotto-Kapelle in Padua," *Sterne und Weltraum* 34 (1995): 278–85.
10. Laura Jacobus, "Giotto's Design of the Arena Chapel," *Apollo* 142, no. 406 (1995): 39–42.
11. Jacobus, "Annunciation," 105.
12. Ibid., 104 (for shutters) and 105 (for gilding).
13. For another account considering the lighting effects throughout the chapel and how they resonate with the overall space see Bellinato, *Giotto*, 9–10 & 38–42.
14. Macfarlane and Martin, *Glass,* 14. For more commentary on this idea see ibid., 3, 40, 51–59.
15. For the most comprehensive treatment of these spaces in terms of the significance of their illusionism see Péter Bokody, *Images-within-Images in Italian Painting (1250–1350): Reality and Reflexivity* (New York and London: Routledge, 2016), 37–58 (First published in 2015 by Ashgate Publishing); Other scholars who have noted and commented on the purpose of these spaces and their successful illusionism include Ursula Schlegel, "On the Picture Program of the Arena Chapel," in *Giotto: The Arena Chapel Frescoes*, ed. James H. Stubblebine (New York and London: Norton, 1969), 196–97; Roberto Longhi, "Giotto spazioso" *Paragone* 31 (1952): 20; Decio Gioseffi, *Giotto architetto* (Milan, 1963), 53; Luciano Bellosi, "La rappresentazione dello spazio," in *Storia dell'arte italiana,* vol. 4 (Turin: Einaudi, 1980): 14–15.
16. Thomas de Wesselow, "The Date of the Saint Francis Cycle," in *The Art of the Franciscan Order in Italy*, ed. William Robert Cook (Leiden: Brill, 2005), 134–42.
17. David Whitehouse, *Medieval Glass for Popes, Princes, and Peasants* (Corning, NY: Corning Museum of Glass, 2010), 53–55 and Daniela Stiaffini, *Il vetro nel medioevo: Tecniche, strutture, manufatti* (Rome: Fratelli Palombi, 1999), 119. For more on glass lamps see Marina Uboldi, "Diffusione delle lampade vitree in età tardoantica e altomedievale e spunti per una tipologia," *Archeologia Medievale* 22 (1995): 93–145, specifically p. 95 for commentary on Giotto's lamps in the Arena Chapel.
18. The similarity in shape can be observed when comparing Giotto's lamps to a later depiction of transparent glass oil lamps painted by Fra Angelico in *Saint Agatha Arising from her Tomb and Appearing to Saint Lucy and her Mother Eustachia* which was formerly in the Feigen Collection. For more on this comparison see Chapter 5 note 8. I'd like to thank Manlio Leo Mezzacasa for this observation regarding the vessels' shape.

19. He would have also encountered numerous passages of illusionism as described by Bokody, *Images-within-Images*, 52–55.
20. This theory is also explored in ibid., 37–58 where Bokody, 37 observes that these passages of illusionism, "see to turn the representation of reality into reality itself, and embedded images seek to differentiate between various registers of perceived reality within the principle images."
21. For his discussion of linear perspective see Samuel Y. Edgerton, *The Renaissance Rediscovery of Linear Perspective* (New York: Basic Books, 1975); Samuel Y. Edgerton, *The Heritage of Giotto's Geometry: Art and Science on the Eve of the Scientific Revolution* (Ithaca, NY: Cornell University Press, 1991); and, most recently, Samuel Y. Edgerton, *The Mirror, the Window, and the Telescope: How Linear Perspective Changed Our Vision of the Universe* (Ithaca, NY: Cornell University Press, 2009).
22. Edgerton suggests that Brunelleschi used a metal mirror in the creation of the piece and a glass mirror in the demonstration of it, see Edgerton, *Mirror*, 50–51. For more on the nature of the demonstration see, ibid., 44–53.
23. Quoted in ibid., 119.

References

Alighieri, Dante. *The Convivio*. Edited and Translated by Andrew Frisardi. Cambridge: Cambridge University Press, 2018.

Bellinato, Claudio. *Giotto: The Scrovegni Chapel*. Ponzano Veneto: Grafiche Vianello Srl., 2006.

Bellosi, Luciano. "La rappresentazione dello spazio." In *Storia dell'arte italiana*, vol. 4, 6–30. Turin: Einaudi, 1980.

Bokody, Péter. *Images-within-Images in Italian Painting (1250–1350): Reality and Reflexivity*. New York and London: Routledge, 2016. (First published in 2015 by Ashgate Publishing).

de Wesselow, Thomas. "The Date of the Saint Francis Cycle." In *The Art of the Franciscan Order in Italy*, edited by William Robert Cook, 134–42. Leiden: Brill, 2005.

Edgerton, Samuel Y. *The Renaissance Rediscovery of Linear Perspective*. New York: Basic Books, 1975.

———. *The Heritage of Giotto's Geometry: Art and Science on the Eve of the Scientific Revolution*. Ithaca, NY: Cornell University Press, 1991.

———. *The Mirror, the Window, and the Telescope: How Linear Renaissance Perspective Changed Our Vision of the Universe*. Ithaca, NY: Cornell University Press, 2009.

Gioseffi, Decio. *Giotto architetto*. Milan, 1963.

Ilardi, Vincent. *Renaissance Vision from Spectacles to Telescopes*. Philadelphia: American Philosophical Society, 2007.

Jacobus, Laura. "Giotto's Design of the Arena Chapel." *Apollo* 142, no. 406 (1995): 37–42.

———. "Giotto's Annunciation in the Arena Chapel, Padua." *The Art Bulletin* 81, no. 1 (March 1999): 93–107.

———. *Giotto and the Arena Chapel: Art, Architecture, and Experience*. Turnhout: Brepols, 2008.

Lesnick, Daniel R. *Preaching in Medieval Florence: The Social World of Franciscan and Dominican Spirituality.* Athens, GA: University of Georgia Press, 1989.

Longhi, Roberto. "Giotto spazioso." *Paragone* 31 (1952): 18–24.

Macfarlane, Alan, and Gerry Martin. *Glass: A World History.* Chicago: University of Chicago Press, 2002.

Romano, Giuliano and Hans Michael Thomas, "Sul significato di alcuni fenomeni solari che si manifestano nella cappella di Giotto a Padova." *Ateneo Veneto* 178 (1991): 213–56.

Schlegel, Ursula. "On the Picture Program of the Arena Chapel." In *Giotto: The Arena Chapel Frescoes*, edited by James H. Stubblebine, 182–202. New York and London: Norton, 1969.

Stiaffini, Daniela. *Il vetro nel medioevo: Tecniche, strutture, manufatti.* Rome: Fratelli Palombi, 1999.

Thomas, Hans Michael. "Sonneneffekte in der Giotto–Kapelle in Padua." *Sterne und Weltraum* 34 (1995): 278–85.

Uboldi, Marina. "Diffusione delle lampade vitree in età tardoantica e altomedievale e spunti per una tipologia." *Archeologia Medievale* 22 (1995): 93–145.

White, John. *Art and Architecture in Italy: 1250–1400.* New Haven, CT: Yale University Press, 1993.

Whitehouse, David. *Medieval Glass for Popes, Princes, and Peasants.* Corning, NY: Corning Museum of Glass, 2010.

Index

Page numbers in italics refer to illustrations.

C

D

H

I

J

K

L

M

N

O

P

R

S

T

U

V

W

www.ingramcontent.com/pod-product-compliance
Ingram Content Group UK Ltd.
Pitfield, Milton Keynes, MK11 3LW, UK
UKHW022207051026
15624UKWH00001B/37

9 781433 148347